Information Extraction and Object Tracking in Digital Video

Edited by Antonio José Ribeiro Neves and Francisco Javier Gallegos-Funes

Published in London, United Kingdom

Information Extraction and Object Tracking in Digital Video
http://dx.doi.org/10.5772/intechopen.94616
Edited by Antonio José Ribeiro Neves and Francisco Javier Gallegos-Funes

Contributors
Vinay M. Malligere Shivanna, Kuan-Chou Chen, Jiun-In Guo, Bo-Xun Wu, Nelli Petrovna Maksymovych, Ludmila Oleksenko, George Fedorenko, Awet Haileslassie Gebrehiwot, Jesus Bescos, Alvaro Garcia-Martin, Soon Ki Jung, Md. Maklachur Rahman, Richa Golash, Yogendra Kumar Jain, Francisco Javier Gallegos-Funes, Miguel Ángel Castillo-Martínez, Guillermo Urriolagoitia-Sosa, Alberto J. Jorge Rosales-Silva, Ronaldo Ferreira, Joaquim José de Castro Ferreira, António J. R. José Ribeiro Neves, Blanca E. Carvajal-Gámez, Izyalith E. Álvarez-Cisneros, David Araujo-Díaz, Laura Méndez Segundo, Miguel A. Castillo-Martínez

First published in London, United Kingdom, 2022 by IntechOpen
IntechOpen is the global imprint of INTECHOPEN LIMITED, registered in England and Wales, registration number: 11086078, 5 Princes Gate Court, London, SW7 2QJ, United Kingdom
Printed in Croatia

British Library Cataloguing-in-Publication Data
A catalogue record for this book is available from the British Library

Additional hard and PDF copies can be obtained from orders@intechopen.com

Information Extraction and Object Tracking in Digital Video
Edited by Antonio José Ribeiro Neves and Francisco Javier Gallegos-Funes
p. cm.

This title is part of the Artificial Intelligence Book Series, Volume 11
Topic: Computer Vision
Series Editor: Andries Engelbrecht
Topic Editor: George Papakostas

Print ISBN 978-1-83969-460-8
Online ISBN 978-1-83969-461-5
eBook (PDF) ISBN 978-1-83969-462-2
ISSN 2633-1403

For EU product safety concerns:
IN TECH d.o.o., Prolaz Marije Krucifikse Kozulić 3, 51000 Rijeka, Croatia,
info@intechopen.com or visit our website at intechopen.com.

IntechOpen Book Series

Artificial Intelligence

Volume 11

Aims and Scope of the Series

Artificial Intelligence (AI) is a rapidly developing multidisciplinary research area that aims to solve increasingly complex problems. In today's highly integrated world, AI promises to become a robust and powerful means for obtaining solutions to previously unsolvable problems. This Series is intended for researchers and students alike interested in this fascinating field and its many applications.

Meet the Series Editor

Andries Engelbrecht received the Masters and Ph.D. degrees in Computer Science from the University of Stellenbosch, South Africa, in 1994 and 1999 respectively. He is currently appointed as the Voigt Chair in Data Science in the Department of Industrial Engineering, with a joint appointment as Professor in the Computer Science Division, Stellenbosch University. Prior to his appointment at Stellenbosch University, he has been at the University of Pretoria, Department of Computer Science (1998-2018), where he was appointed as South Africa Research Chair in Artifical Intelligence (2007-2018), the head of the Department of Computer Science (2008-2017), and Director of the Institute for Big Data and Data Science (2017-2018). In addition to a number of research articles, he has written two books, Computational Intelligence: An Introduction and Fundamentals of Computational Swarm Intelligence.

Meet the Volume Editors

Prof. António J. R. Neves received a Ph.D. in Electrical Engineering from the University of Aveiro, Portugal, in 2007. Since 2002, he has been a researcher at the Institute of Electronics and Informatics Engineering of Aveiro. Since 2007, he has been an assistant professor in the Department of Electronics, Telecommunications, and Informatics, University of Aveiro. He is the director of the undergraduate course on Electrical and Computers Engineering and the vice-director of the master's degree in Electronics and Telecommunications Engineering. He is an IEEE Senior Member and a member of several other research organizations worldwide. His main research interests are computer vision, intelligent systems, robotics, and image and video processing. He has participated in or coordinated several research projects and received more than thirty-five awards. He has 161 publications to his credit, including books, book chapters, journal articles, and conference papers. He has vast experience as a reviewer of several journals and conferences. As a professor, Dr. Neves has supervised several Ph.D. and master's students and was involved in more than twenty-five different courses.

Francisco J. Gallegos-Funes received his Ph.D. in Communications and Electronics from the Instituto Politécnico Nacional de México (National Polytechnic Institute of Mexico) in 2003. He is currently an associate professor in the Escuela Superior de Ingeniería Mecánica y Eléctrica (Mechanical and Electrical Engineering Higher School) at the same institute. His areas of scientific interest are signal and image processing, filtering, steganography, segmentation, pattern recognition, biomedical signal processing, sensors, and real-time applications.

Contents

Preface

The research on computer vision systems has been increasing every day and has led to the design of multiple types of these systems with innumerous applications in our daily life. The recent advances in artificial intelligence (AI), together with the huge amount of digital visual data available, have boosted vision system performance in several ways. Computer systems with digital cameras using the most recent AI techniques are intelligent to perceive and understand the visual world. However, there are still several open challenges in using computer vision at the level of the human eye, despite the existence of several systems that outperform humans in several specific tasks.

Information extraction and visual object tracking are essential tasks in computer vision with a huge number of real-world applications including intelligent management of Web videos, events and object detection, human-computer interaction, augmented reality, autonomous vehicles, robotics, sports, video indexing and retrieval, and surveillance and security, among many others. The goal of information extraction is to efficiently extract useful data from the images recorded over time in the form of digital videos. Visual object tracking aims to estimate an unknown visual trajectory of a target when an initial position is given in a video frame. The exponential effort in the development of new algorithms for information extraction and visual object tracking in digital videos is confirmed by the amount of money invested in scientific projects and in the quantity and quality of the manuscripts published in a considerable number of journals and conferences worldwide on these topics.

This book is a result of research done by several researchers and professionals who have highly contributed to the field of image processing. The main goal is to present recent advances in this hot topic. I would like to thank all the authors for their excellent contributions.

This book contains eight chapters divided into three sections. Section 1 consists of four chapters focusing on the visual tracking problem. Section 2 includes three chapters focusing on information extraction from images. Section 3 includes one chapter on new advances in image sensors.

I hope that readers of this book will find it interesting and informative, considering it a good tool for their research or projects.

António José Ribeiro Neves, Ph.D.
Professor,
Department of Electronics, Telecommunications, and Informatics,
University of Aveiro,
Aveiro, Portugal

Francisco Javier Gallegos-Funes, Ph.D.
Instituto Politécnico Nacional,
Mexico City, Mexico

Section 1

Object Tracking

Chapter 1

Object Tracking Using Adapted Optical Flow

Ronaldo Ferreira, Joaquim José de Castro Ferreira and António José Ribeiro Neves

Abstract

The objective of this work is to present an object tracking algorithm developed from the combination of random tree techniques and optical flow adapted in terms of Gaussian curvature. This allows you to define a minimum surface limited by the contour of a two-dimensional image, which must or should not contain a minimum amount of optical flow vector associated with the movement of an object. The random tree will have the purpose of verifying the existence of superfluous vectors of optical flow by discarding them, defining a minimum number of vectors that characterizes the movement of the object. The results obtained were compared with those of the Lucas-Kanade algorithms with and without Gaussian filter, Horn and Schunk and Farneback. The items evaluated were precision and processing time, which made it possible to validate the results, despite the distinct nature between the algorithms. They were like those obtained in Lucas and Kanade with or without Gaussian filter, the Horn and Schunk, and better in relation to Farneback. This work allows analyzing the optical flow over small regions in an optimal way in relation to precision (and computational cost), enabling its application to area, such as cardiology, in the prediction of infarction.

Keywords: Object tracking, vehicle tracking, optical flow, gaussian curvature, random forest

1. Introduction

Object tracking is defined as a problem of estimating the object's trajectory, done by means of a video image. There are several tools for tracking objects and are used in various fields of research, such as computer vision, digital video processing, and autonomous vehicle navigation [1]. With the emergence of high-performance computers, high-resolution cameras, and the growing use of so-called autonomous systems that, in addition to these items, require specialized tracking algorithms, increasingly accurate and robust for automatic video analysis, has currently been the target of numerous research on the development of new object tracking techniques [2, 3].

Object tracking techniques are applicable to motion-based reconnaissance cases [4], automatic surveillance systems [5], pedestrian flow monitoring in crosswalks [6], traffic control [7], and autonomous vehicular navigation [8]. Problems of this type are highly complex due to the characteristics of the object and the environment, generating many variables, which impairs performance and makes the application of tracking algorithms unfeasible to real-world situations. Some approaches seek to resolve this impasse by simplifying the problem, reducing the number of variables [9]. This process, in most cases, does not generate good results [10, 11], making it even more difficult to identify the main attributes to be selected to perform a task [12, 13].

Most of the object tracking problems occur in open environments, so-called uncontrolled [14]. The complexity of these problems has attracted the interest of the scientific community and generated numerous applied research in various fields of research. Current approaches, such as the ones that use convolutional neural networks —CNN, deal well with the high number of variables of these types of problems, providing space–temporal information of the tracked objects, through three-dimensional convolutions [15–17]. This ends up creating an enormous number of learnable parameters, which ends up generating an overfitting [11]. A solution to reduce this number of learnable parameters was combining space–time data, extracted using the optical flow algorithm, used in the Two-Stream technique [18–20]. However, this technique presents good results only for large datasets, showing itself to be inefficient for small datasets [15, 21].

In recent years, research using machine learning has been applied to tracking problems, gaining notoriety due to the excellent results obtained in complex environments and attribute extraction [21–23]. Deep learning stands out among these techniques for presenting excellent results to unsupervised learning problems, [24], object identification [25], semantic segmentation [26]. Random trees are also examples of machine learning techniques, and their excellent results, due to their precision and great capacity to handle a large volume of data and low overfitting tendency [27, 28], and widely used in research areas such as medicine, in the prediction of hereditary diseases [29], agriculture to increase the productivity of a given plantation crop and in astronomy, acting on the improvement of images captured by telescopes, in the spectrum electromagnetic radiation not visible to the human eye [30]. The possibilities of applications, and new trends and research related to machine learning techniques, with particular attention to random trees, allow the development of algorithms that can be combined with existing ones, in the case of optical flow algorithms, (belonging to computational field of view) taken advantage of in this way, the advantages of each [31–33].

Developing an algorithm whose objective is to track objects, using the particular advantages of these techniques in a combined way, justifies creating a tracking algorithm that combines the optical flow technique, adapted in this work in terms of the Gaussian curvature associated with a minimal surface, with a random trees waiting for it to capture on this surface a minimum number of optical flow vectors that characterize the moving object, accurately and with low computational cost, contributing not only in the fields of computational vision but in other branches of science, such as in medicine, it can help in the early identification of infarctions.

2. Related works

Due to the large number of studies related to the technique of object tracking, only a small number surrounding this theme will be addressed. The focus of this project is not to make a thorough study on the state of the art. With this in this item, the main works in the literature, associated with the tracking of objects, will be presented. Among the various approaches used for this context, we highlight those focused on the techniques of optical flow, and others belonging to machine learning, such as those that use identifications of patterns, which allow relating, framing, and justifying the development of this proposal and its importance, through its contribution, to the state of the art.

2.1 Object tracking

Object tracking is defined as a process that allows you to uniquely estimate and associate the movements of objects with consecutive image frames. The objects considered can be from one, the set of pixels belonging to a region of the image.
The detection of pixels is done by a motion detector or objects, which allows to locate objects with similar characteristics that move, between consecutive frames.

These characteristics of the object to be tracked are compared with the characteristics of a reference object modeled by a classifier over a limited region of the so-called region of interest frame, where the probability of detection of the object is greater. Thus, according to [33], the detector of traced objects, locate several objects on the different parts of the region of interest and performs the comparison of these objects with the reference object. This process is performed for each frame and each object detected, candidate to be recognized as the greatest possible similarity, to the reference object can be represented, through a set of fixed-size characteristics, extracted from this region containing a set of pixels, which can be represented by a numerical array of data.

Thus, mathematically, the region containing a set of pixels belonging to the regions of the object of interest, where the characteristics that allow to test whether the region of the frame, in which the object to be traced is, is given by:

$$\mathrm{OC_i(t) = OC(t)|||L(OC(t)) - L(OR(i-1))|| < \varepsilon} \quad (1)$$

where, $L(OC(t))$ is the position (x, y) of the centroid of the candidate object $OC(t)$, $L(OR(i-1))$, is the position of the object traced to the (i–1)—frame of the video and $0 < \varepsilon \in \mathbb{R}$, is an actual value associated with the size of the region of the object of interest.

According to the works of [34, 35], learning methods are used to adapt the changes of movement and other characteristics such as geometric aspect and appearance of the tracked object. These methods are usually used adaptive tracked object trackers and detectors. The following will be presented other types of object trackers, found in the literature.

According to [36], a classifier can be defined with a f belonging to a family of functions F parameterized by a set of classifier parameters. They form a detector of objects to be tracked which in turn is an integral part of a tracker. A classifier can also be training and thereby generate a set of classification parameters, producing the function$f$, that allows you to efficiently indicate the classes v_i of the test data x_i from a training set $C_t = \{(x_1, y_1), \ldots, (x_n, y_n)\}$. The data is points in *the space of* the characteristics, which can be entropy, the gray level, among others.

The classifier aims to determine the best way to discriminate the data classes, on the space of characteristics. The test data form a set containing the characteristics of the candidate objects, which have not yet been classified. The position of the object to be tracked in the frame is defined as the position corresponding to the highest response of the detector of the object to be tracked on the *ith*-candidate objects. Therefore, the position of the object to be tracked is determined by the position of the *ith*-candidate object, which is most likely to belong to the class of the crawled object, given by the following equation:

$$L(OR(t)) = L(\text{argmaxiP}(yi = COR|OCi(t), PR(t))) \quad (2)$$

$$P(yi = COR|OCi(t), PR(t)) = \{ 1/N, \text{se}\|L(OCi\ (t)) - L(OR(t|t-1))\| < \varepsilon\ 0, \text{otherwise} \quad (3)$$

where the variable *COR* in the equation (3), are the classes of the tracked objects and the candidate objects *OC*, all with equiprobability of occurrence. According to the types of classifiers used object detectors to be tracked, along with the initial detector, it is possible to use some learning technique to train them. One of the ways used is offline training [36] and adjusting the parameters of the classifier before running the tracker.

Offline-trained classifiers are generally employed in object detectors designed to detect all new objects of interest that enter the camera's field of view [37]. The training set C_t, must contain characteristics x_i extracted from the objects to be traced and diverse environmental characteristics. This allows new objects, with varied geometric characteristics and aspects, to be detected more efficiently. As for online training, the adjustment of the parameters of the classified is performed during the tracking process. For online trained classifiers, they are generally used in object detectors to be tracked. Thus, in each frame, the new extracted characteristics are used to adjust the classifiers.

2.1.1 Binary classification

In [38], trackers that use the detection tracking technique deal with object tracking, as a binary classification problem whose goal is to find the best function f that separates the objects to be tracked *R*, of other objects in the environment. Object tracking seen as a binary classification problem, which is currently one of the subjects that receives the most attention in research in computing vision.

In [39], were developed trackers that used detectors of objects to be tracked, formed by classifiers in committee formed by binary classifiers said weak. For [40], a binary classifier is defined as a classifier, used in problems where the class y_i of a OC_i belongs to the set $Y = \{-1, 1\}$. The negative class $\{-1\}$ refers to the characteristics of the environment and other objects. A positive class {+1} refers to the class of the object to be tracked.

A classifier is said to be weak, when it has a probability of "hitting" a given data class, only slightly higher than a random classifier. The detector of the object to be tracked must separate the crawled object from the other objects and the environment. Its purpose and determine the position of the tracked object, according to the equations (1)–(3). According to [41, 42] each of the *ith* candidate object classes OC_i, it is defined according to Bayesian theory of decision, through minimal classification error. This means that, the decision given by observing, the sign of the difference between $P(y_i = COR|OC_i(t), PR(t))$ and $P(y_i = CNOR|OC_i(t), PR(t))$, so that the sum of these probabilities is unitary.

2.1.2 Monitoring systems

For [43], the term monitoring system, refers to the process of monitoring and autonomous control, without human intervention. This type of system has the function of detecting, classifying, tracking, analyzing, and interpreting the behavior of objects of interest. In [44, 45], this technique was used combined with statistical techniques for controlling people's access to a specific location. It was also observed the use of intelligent monitoring systems, applied to building, port, or ship security [46, 47].

The functions comprised by a monitoring system are so-called low- and high-level tasks. Among some high-level tasks, we highlight the analysis, interpretation and description of behavior, the recognition of gestures, and the decision between the occurrence or not of a threat. Performing high-level tasks require that for each frame, the system needs to perform low-level tasks, which involve direct manipulation of the image pixels [48–56]. As an example, we highlight the processes of noise elimination, detection of connected components, and obtain information on the location and geometric aspect of the object of interest.

A monitoring system consists of five main components, which are presented in **Figures 9**. Some monitoring systems may not contain all components. The initial detector aims to detect the pixel regions of each frame that have a significant probability of containing an object to be tracked. This detector can be formed by a motion detector that detects all moving objects based on models of objects previously recorded in a database or based on characteristics extracted offline [40, 41]. The information obtained by the initial detector is processed by an image processor], which will have the function of eliminating noise, segmenting, and detecting the connected components.

The regions containing the most relevant pixels are analyzed and then classified as objects of interest by the classifier [50–54]. Objects of interest are modeled and are now called reference objects so that the tracker determines its position frame by frame [55, 56]. The information obtained by the initial detector is processed by an image processor], which will have the function of eliminating noise, segmenting, and detecting the connected components.

A tracker, an integral part of a detector, is defined as a function that allows estimating the position of objects at each consecutive frame, through and defines the region of the object of interest, for each ith object being tracked within a region of interest. This estimation of the movement is performed through the correct association of the captured and tracked objects, to consecutive video frames. The trace often and interpreted as a data binding problem. **Figure 1** shows a schematic of the main components of a monitoring system.

2.2 State of the art in object tracking with optical flow

Several techniques that allow the calculation to have been developed in recent years to calculate the optical flow vector [57]. These methods are grouped according to

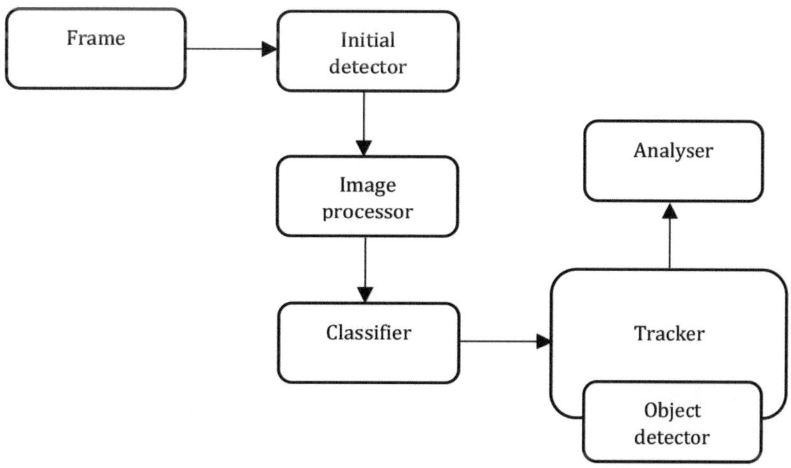

Figure 1.
Main component of a monitoring.

their main characteristics and the approach used for the calculation of the optical flow. Thus, the differential methods performed in the studies in [56], the methods d and calculation of the optical flow through the frequency domain [46] the phase correlation methods [58], and the method of association between regions [59].

The method proposed in [56], allows the calculation of the optical flow for each point around a neighborhood of pixels. In [60], it is also considered a neighborhood of pixels, but in this case, the calculation of the optical flow is performed geometrically. In the work presented by [61] it is adding of the restrictions of regularization. In [62] turn active compare performance analyses were performed between the various algorithms and optical flow present in the literature.

This technique is considered robust for detaining and tracking moving objects from your images, both those captured by fixed or mobile cameras. This gives this technique, but high computational cost makes most practical applications unfeasible. Thus, to reduce this complexity, techniques of increasing resolutions were adopted in [63]. Also, for the same purpose, we used the techniques of subsampling on some of the pixels belonging to the object of interest to obtain optical flow [52].

Other authors also use a point of interest detector to select the best pixels for tracking and calculate the optical flow on these points [52, 64]. The reduction in the number of points to be tracked is associated with a decrease in computational complexity, so in [52] the points of interest were selected using the FAST algorithm [64].

The method developed by Lucas-Kanade [56], it is a differential method and widely used in the literature and having variations modifications. It allows you to estimate the optical flow for each point (x, y, t) calculating the like transformation $(TA\ (x, t)$, applied to the pixels of a pixel grid, with center in (x, y) by the following function $f(x, t)$, that is:

$$f(x,t) = \min\left(\sum_{x \in (pixel\ grid)} [Q(x, t-1) - (Q(TA(x), t)] * g(x)]\right) \tag{4}$$

where $g(x)$ is a Gaussian smoothing filter centered on x.

New variations of the techniques were being proposed to make the calculation of the optical flow faster and faster. In [65] a tracker was proposed based on the algorithm of [56]. The translation of a point represented by a grid of rectangular sized pixels 25×25, was calculated and its validity is evaluated by calculating the SSD[1] in the grid pixels in $Q(t)$ and in $Q(t-1)$. If the SSD is high, the point is dropped and stops being traced.

In [51] objects were detected by subtracting the image from the environment and removed the movement of the camera with the calculation algorithm of the optical flow vector proposed by [56]. In the studies carried out in [66, 67], they showed that the reliability of the estimated optical flow reduced the case of some points of the object of interest whose optical flow cannot be represented by the same matrix given by the related transformation $(TA\ (x, t))$ of the other points. Thus, to improve the robustness of the algorithm of [56, 67] proposed a calculation of the independent optical flow vector for each of the N points belonging to the object of interest selected with the SURF (Speeded Up Robust Features) point detector in the initial frame.

In [67] they also modified Lucas - Kanade's algorithm [56] by inserting the Hessian matrix in the calculation of the value of the variation of the related transformation

[1] Actual amount of data rate for actual data recorded.

($\Delta TA\ (x, t)$). The algorithm allows for more effective tracking when partial occlusions, deformations, and changes in lighting occur, as optical flow is not calculated considering all points of objects of interest.

Already in the proposal presented in [68] was the development of algorithm to detect people in infrared images that combines the information of the value of pixels with a method of motion detection. The algorithm forms a relevant pixel map by applying thresholding segmentation. While the camera is still, an image M is built with the differentiation between frames. If the camera is in motion, M is filled with the pixels obtained by the analysis of the moment of the optical flow calculated by the algorithm of [56]. The map of relevant pixels is replaced by the union between M and the Pixel Map relevant to the first case and the second an interception between M and the pixel map relevant case to compensate for the movement of the camera.

The method for tracking swimmers presented in [46], uses the information of the movement pattern by the optical flow and the appearance of the water that is modeled by a MoG.[2] This allows you to calculate an optical flow vector for each pixel of the video independently of the other, through B which is an array composed of gradients in the directions x and y pixels in a grid of pixels.

In [69], a method was presented that incorporated physical restrictions to the calculation of optical flow. The tracker uses the constraints to extract the moving pixels with a lower failure rate. The calculation can be impaired when occlusions occur or when the environment has low light. The operator defines the physical constraints and selects the points of the *OR* that are tracked by optical flow. Constraints can be geometric, kinematic, dynamic, of the property of the material that makes up the *OR* or any other type of restriction.

In [70], the points that are tracked with the optical flow are defined by applying the Canny edge detector on the pixels of the reference pixel map. Pixels that produce a high response to the Canny detector are the selected points.

In [43], optical flow is used as a characteristic for tracking the contour of the object. The contour is shifted in small steps until the position in which the optical flow vectors are homogeneous is found.

In [64], they performed an estimate of the translation and orientation of the reference object by calculating the optical flow of the pixels belonging to its silhouette. The coordinates of the centroid position are defined by minimizing the Hausdorff distance between the mean of the optical flow vectors of the reference object and the candidate object to be chosen as the object of interest.

2.2.1 Optical flow as a function of Gaussian curvature

Optical flow is defined as a dense vector field associated with the movement and apparent velocity of an object, given by the translation of pixels from consecutive frames in an image region. It can be calculated from the brightness restriction, considered constant, from the corresponding pixels in consecutive frames.

Mathematically be a pixel (x, y), associated with a luminous intensity $I(x, y)$, over an image surface or plane, and a time interval and a sequence of frames associated with an apparent offset of the pixel over that image surface or plane. Thus, the rate of variation of light intensity in relation to a time interval, associated with the apparent movement of the pixel, on a surface or plane of the image, being considered practically null can be given by:

---

[2] MoG: mixture of Gaussian distributions.

$$\frac{dI(x,y)}{dt} = \frac{\partial I(x,y)}{\partial x}\frac{d\,x}{dt} + \frac{\partial I(x,y)}{\partial y}\frac{d\,y}{dt} + \frac{\partial I(u,v)}{dt} \tag{5}$$

$$= I_x\, u_x + I_y\, v_x + I_t \tag{6}$$

$$\frac{dI(x,y)}{dt} = 0 \Longrightarrow I_x\, u_x + I_y\, v_x + I_t = 0 \tag{7}$$

So that equation (7) is called optical flow restriction and where the terms I_x, I_y, I_t denote the derivatives relative to the brightness intensity relative to the coordinates x, y and time t, and u and v, $(u(x,y), v(x,y))$ are the horizontal and vertical components of a vector representing the optical flow field, for the pixel (x, y) in question.

The number of variables in equation (6) is greater than that of equations, which does not allow estimating components and vector, and determining a single solution for the optical flow restriction equation. With this, Lucas and Kanade proposed a solution to solve this problem. The solution method proposed by them considers the constant flow in a region formed by a set of pixels $N \times N$, so you can write the optical flow restriction equation for each pixel in this region, thus obtaining a systems of equations with 2 variables, that is:

$$\begin{aligned} \mathrm{I}_{x1}\mathrm{v}_x + \mathrm{I}_{y1}\mathrm{v}_y + \mathrm{I}_{t1} &= 0 \\ \mathrm{I}_{x2}\mathrm{v}_x + \mathrm{I}_{y2}\mathrm{v}_y + \mathrm{I}_{t2} &= 0 \\ &\vdots \\ \mathrm{I}_{xp}\mathrm{v}_x + \mathrm{I}_{yp}\mathrm{v}_y + \mathrm{I}_{tp} &= 0 \end{aligned} \tag{8}$$

Passing the set of equations given by equation (8) to the matrix form we have:

$$\begin{pmatrix} \mathrm{I}_{x1} & \mathrm{I}_{y1} \\ \vdots & \vdots \\ \mathrm{I}_{xp} & \mathrm{I}_{yp} \end{pmatrix} \begin{pmatrix} \mathrm{v}_x \\ \mathrm{v}_y \end{pmatrix} + \begin{pmatrix} \mathrm{I}_{t1} \\ \vdots \\ \mathrm{I}_{tp} \end{pmatrix} = \begin{pmatrix} 0 \\ \vdots \\ 0 \end{pmatrix} \tag{9}$$

Using the least squares method, in the system of equations (9) in the form of matricial, the same can be solved. Therefore, the optical flow $v = (v_x, v_y)$ can be estimated for a particular region or window with $N \times N$ pixels, that is:

$$\begin{aligned} &\begin{pmatrix} \mathrm{I}_{x1} & \mathrm{I}_{y1} \\ \vdots & \vdots \\ \mathrm{I}_{xp} & \mathrm{I}_{yp} \end{pmatrix} \begin{pmatrix} \mathrm{I}_{x1} & \mathrm{I}_{y1} \\ \vdots & \vdots \\ \mathrm{I}_{xp} & \mathrm{I}_{yp} \end{pmatrix}^{t} \begin{pmatrix} \mathrm{v}_x \\ \mathrm{v}_y \end{pmatrix} = -\begin{pmatrix} \mathrm{I}_{t1} \\ \vdots \\ \mathrm{I}_{tp} \end{pmatrix} \begin{pmatrix} \mathrm{I}_{x1} & \mathrm{I}_{y1} \\ \vdots & \vdots \\ \mathrm{I}_{xp} & \mathrm{I}_{yp} \end{pmatrix}^{t} \\ &\Rightarrow \begin{pmatrix} \mathrm{I}_{x1} & \mathrm{I}_{y1} \\ \vdots & \vdots \\ \mathrm{I}_{xp} & \mathrm{I}_{yp} \end{pmatrix} \begin{pmatrix} \mathrm{I}_{x1} & \mathrm{I}_{y1} \\ \vdots & \vdots \\ \mathrm{I}_{xp} & \mathrm{I}_{yp} \end{pmatrix}^{t} \left(\begin{pmatrix} \mathrm{I}_{x1} & \mathrm{I}_{y1} \\ \vdots & \vdots \\ \mathrm{I}_{xp} & \mathrm{I}_{yp} \end{pmatrix} \begin{pmatrix} \mathrm{I}_{x1} & \mathrm{I}_{y1} \\ \vdots & \vdots \\ \mathrm{I}_{xp} & \mathrm{I}_{yp} \end{pmatrix}^{t} \right)^{-1} \begin{pmatrix} \mathrm{v}_x \\ \mathrm{v}_y \end{pmatrix} \\ &= -\begin{pmatrix} I_{t1} \\ \vdots \\ I_{tp} \end{pmatrix} \begin{pmatrix} I_{x1} & I_{y1} \\ \vdots & \vdots \\ I_{xp} & I_{yp} \end{pmatrix}^{t} \left(\begin{pmatrix} I_{x1} & I_{y1} \\ \vdots & \vdots \\ I_{xp} & I_{yp} \end{pmatrix} \begin{pmatrix} I_{x1} & I_{y1} \\ \vdots & \vdots \\ I_{xp} & I_{yp} \end{pmatrix}^{t} \right)^{-1} \end{aligned} \tag{10}$$

Where:

$$\begin{pmatrix} I_{x1} & I_{y1} \\ \vdots & \vdots \\ I_{xp} & I_{yp} \end{pmatrix} = A_{px2} = A$$

Therefore, one has that:

$$\begin{pmatrix} I_{x1} & I_{y1} \\ \vdots & \vdots \\ I_{xp} & I_{yp} \end{pmatrix} \begin{pmatrix} I_{x1} & I_{y1} \\ \vdots & \vdots \\ I_{xp} & I_{yp} \end{pmatrix}^t \left(\begin{pmatrix} I_{x1} & I_{y1} \\ \vdots & \vdots \\ I_{xp} & I_{yp} \end{pmatrix} \begin{pmatrix} I_{x1} & I_{y1} \\ \vdots & \vdots \\ I_{xp} & I_{yp} \end{pmatrix}^t \right)^{-1} =$$

$$= (A \cdot A^t)(A \cdot A^t)^{-1} = \begin{pmatrix} 1 & \cdots & 0 \\ \vdots & \ddots & \vdots \\ 0 & \cdots & 1 \end{pmatrix} = Id_{pxp} = Id \tag{11}$$

Thus:

$$Id = -I_t \cdot A^t (A \cdot A^t)^{-1} \tag{12}$$

This method has a reduced computational cost to determine optical flow estimation when compared to other methods because it is simple, that is, it is since the region in which the variation of light intensity between pixels is minimal has a size 2×2, contained in a region *N x N*. In this way, the Optical Flow is determined in a region of 2×2, between these two pixels, using only one matrix inversion operation (equation (12)).

To calculate the optical flow over the size region *N x N*, partial derivatives must be calculated in each pixel. However, considering almost null the variation of the intensity of light between pixels, over the region, the small differences in the accumulated intensities of brightness between pixels compromise the accuracy of the Optical Flow in relation to the determination of the actual motion object, that is, it gains in the processing speed and loses precision in the determination of the motion. When deriving equation (5) we have equation (13), that is:

$$\xi_\alpha^2 = a_x \alpha_1 + a_y \alpha_2 + a_3 u^2 + a_4 v^2 + a_5 uv + a_6 u + a_7 v + a_8 \tag{13}$$

Where the terms $\alpha_x = \frac{\partial v_x}{\partial t}$, $\alpha_y = \frac{\partial v_y}{\partial t}$, are called the components of the acceleration vector, (v_x, v_y) is the components of the velocity vector and the terms $a_1 = I_x = \frac{\partial}{\partial x} I(x,y,t); a_2 = I_y = \frac{\partial}{\partial y} I(x,y,t); a_3 = I_{xx} = \frac{\partial^2}{\partial x^2} I(x,y,t); a_4 = I_{yy} = \frac{\partial^2}{\partial x^2}(x,y,t); a_5 = I_{xy} = \frac{\partial}{\partial xy} I(x,y,t); a_6 = I_{xt} = \frac{\partial}{\partial xt} I(x,y,t); a_7 = I_{yt} = \frac{\partial}{\partial yt} I(x,y,t)\ a_8 = I_{tt} = \frac{\partial}{\partial t^2} I(x,y,t)$, are the first and second partial derivatives of the $I(x,y,t)$.

In view of the small variations present and accumulated along the vector field associated with the optic flow, which cause an additional error in equation (13), a regularization adjustment was made, given by equation (14):

$$\xi_c^2 = \left(\frac{\partial}{\partial x} v_x(x,y)\right)^2 + \left(\frac{\partial}{\partial y} v_x(x,y)\right)^2 + \left(\frac{\partial}{\partial y} v_y(x,y)\right)^2 + \left(\frac{\partial}{\partial x} v_y(x,y)\right)^2 \tag{14}$$

Thus, combining equations (13) and (14), the error ξ can be minimize by the equation (15):

$$\iint\left(\xi_{\alpha}^{2}+\alpha^{2}\xi_{c}^{2}\right)dxdy \tag{15}$$

where α is the value of the weights required for smoothing the variation of the associated optical flow. So, to get the $v_x = v_x(x,y)$ e $v_y = v_y(x,y)$, thus using the resources of the variational calculation one has:

$$\begin{cases} 2\alpha_3 v_x + \alpha_5 v_y = \alpha^2\nabla^2 v_x - b_1 \\ \alpha_5 v_x + 2\alpha_4 v_y = \alpha^2\nabla^2 v_y - b_2 \end{cases} \tag{16}$$

where $\nabla^2 v_x$ is the Laplacian of v_x e $\nabla^2 v_y$ is the Laplacian of v_y and the coefficients de b_1, b_2 can be given as:

$$\begin{cases} b_1 = \dfrac{\partial}{\partial t}\alpha_1 + \alpha_6 \\ b_2 = \dfrac{\partial}{\partial t}\alpha_2 + \alpha_7 \end{cases} \tag{17}$$

and replacing the coefficients I_x, I_y, I_{xx}, I_{yy}, I_{xy}, I_{xt}, I_{yt}, I_{tt} in equation (16), one has:

$$I_{xx}v_x + I_{xy}v_y = \left(\frac{\alpha^2}{3}\right)\nabla 2v_y - I_{xt} \tag{18}$$

$$I_{xy}v_x + 2I_{xy}v_y = \left(\frac{\alpha^2}{3}\right)\nabla 2v_x - I_{yt} \tag{19}$$

whereas $\nabla^2 v_x = \overline{v_{x_1 j_k}} - v_{x_1 j_k}$, e $\nabla^2 \mathrm{v_y} = \overline{v_{y_1 j_k}} - v_{j_1 j_k}$, are the Laplacians of equations (18) and (19), given in their discretized digital forms together with equation (20),

$$\lambda = \left(\frac{\alpha^2}{3}\right) \tag{20}$$

It is possible to reduce the data system by (17), such as:

$$\left[\lambda^2\left(I_{xx} + I_{yy} + \lambda^2\right) + \kappa\right]v_x = \lambda^2\left(I_{yy} + \lambda^2\right)\overline{v_x} - \lambda^2 I_{xy} + \overline{v_x} + c_1 \tag{21}$$

$$\left[\lambda^2\left(I_{xx} + I_{yy} + \lambda^2\right) + \kappa\right]v_x = -\lambda^2\left(I_{yy}\overline{v_x} + \lambda^2\left(I_{yy} + \lambda^2\right)\overline{v_y} + c_2 \tag{22}$$

where the term $\kappa = I_x I_{yy} - {I_{xy}}^2$, it is called *Gaussian curvature* of the surface. And it is also that:

$$\begin{cases} c_1 = I_{xy}I_{yt} - I_{xx}\left(I_{yy} + \lambda^2\right) \\ c_2 = I_{xx}I_{yy} - I_{xx}\left(I_{xx} + \lambda^2\right) \end{cases} \tag{23}$$

Where c_1, c_2 they're real constants.

Therefore, isolating terms v_x, v_y and still replacing c_1 e c_2 in equations (21) and (22) respectively, resulting in equations (24) and (25):

$$v_x = \overline{v_x} - \left[\frac{\lambda^2 (I_{yy} + \lambda^2) \overline{v_x} - \lambda^2 I_{xy} + \overline{v_x} + c_1}{\left[\lambda^2 (I_{xx} + I_{yy} + \lambda^2) + \kappa\right] v_x} \right] \tag{24}$$

$$v_y = \overline{v_y} - \left[\frac{-\lambda^2 (I_{yy} \overline{v_x} + \lambda^2 (I_{yy} + \lambda^2) \overline{v_y} + c_2}{\left[\lambda^2 (I_{xx} + I_{yy} + \lambda^2) + \kappa\right] v_x} \right] \tag{25}$$

The Algorithm 1 is a pseudocode to generate the proposed optical flow vector, through equations (24) and (25) and that allow estimating the speed and position of an object, through a sequence of video images.

Algorithm 1. Adapted optical flow (Gaussian curvature κ).

Begin
Input: Image sequence (video)
Output: Vector optic flow generator (v_x, v_y)
    For I = 1...N do

        Convert images to a gray tone
        Calculate the partial derivatives of 1°and 2°orders of $I(x, y, t)$
        Calculate constants $a_x, a_y, a_1, \ldots, a_8, b_1, b_2, \lambda, c_1, c_2$
        Calculate the discretized Laplacians of $\nabla^2 v_x, \nabla^2 v_y$
        Calculating Gaussian curvature κ
        Calculate flow components (u, v)

    End For

End

2.2.2 Random forests

Developed by Breiman [63] in the mid-2000s, and later revised in [71] random trees are considered one of the best-supervised learning methods used in data prediction and classification. Due to its simplicity, low computational cost, great potential to deal with a large volume of data, and still present great accuracy of results, currently this method has become very popular being applied in various fields of science as data science [72]. Bioinformatics, Ecology, in real-life systems and recognition of 3D objects. In recent years, several studies have been conducted with the objective of making the technique more elaborate and seeking new practical applications [73–75].

Many studies were carried out with the aim of narrowing the existing gap between theory and practice can be seen in [58, 76–78]. Among the main components of random tree forests, one can highlight the bagging method [63], and the criterion of classification and regression called cart-*split* [79], which play critical roles.

Bagging (a bootstrap-aggregating contraction) is an aggregation scheme, which generates samples through the bootstrap method, from the original dataset. These methods are nonparametric and belong to the Monte Carlos method class [80], treating the sample as a finite population. Still, these methods are used when the distribution of the target population is not specified, and the sample is the only information available. How in this way a predictor of each sample is constructed, so that the decision is made through an average, and is more effective computational procedures to improve the indexable estimates, especially for large sets of high-dimensional data, where finding a good model in one step is impossible due to the complexity and scale of the problem. As for the cart-split criterion, it originates from the CART program [63], and is used in the construction of individual trees to choose the best cuts perpendicular to the Axes. However, while bagging and the CART division scheme are key elements in the random forest, both are difficult to mathematically analyze and are a very promising field for both theoretical and practical research.

In general, the set of trees is organized in the form of $\{T_1\ (\Theta_1), T_2\ (\Theta_2) \ldots T_i\ (\Theta_i\}$, where T_B is every tree and Θ_B are bootstrap samples with spare dimensions q x *mtry*, where *mtry* is equal to the number of variables that will be used on each node during the construction of each tree and q is approximately $0,67 \times (n)$. Each of the trees produces a response $y_{1,i}$ for each of the samples W $\{T_1(W) = y_{1,i}, T_2(W) = y_{2,i}, .., T_i(W) = y_{2,B}\}$ and the mean (regression) or majority vote (classification) of the tree responses will be the final response of the model for each of the samples.

3. Methodology

The methodology employed consisted of combining the optical flow algorithm in terms of Gaussian curvature, developed in this work together with the technique of random forest. The language used for the development of this algorithm was the MATLAB programming language, executed on a 64-bit 8th generation notebook, CORE i7. The input data is a video extension Avi, lasting 5 min of a vehicle and two cyclists, circulating in the vicinity of the beach of Costa Nova, in the locality of Ilhavo, in Aveiro, Portugal. The video was fragmented into a set of frames, analyzed two by two by the algorithm for the generation of the vector field of optical flow. After that, the resulting image associated with the flow and a minimal surface region, given by the Gaussian curvature. Next on this surface, the random trees analyzed which vectors presented important characteristics to characterize in an "optimal" way, the movement of the object (see **Figure 2**).

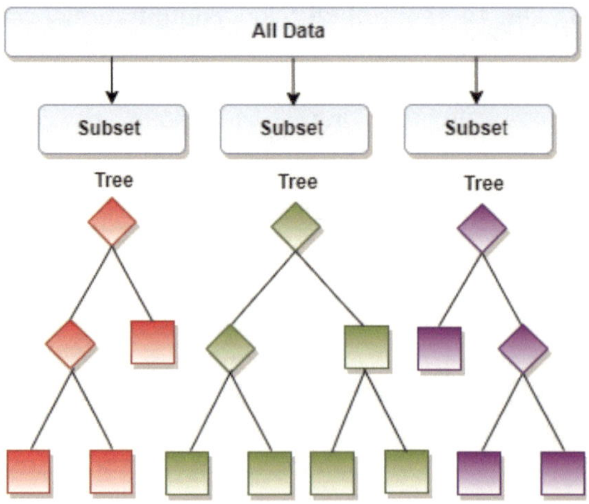

Figure 2.
Representative model of operation of a random forest.

After finishing the process of analysis of the movement of the objects, the execution times and accuracy of the results obtained by the proposed algorithm were compared in relation to the algorithms of Lucas Kanade, Horn and Shunck, Farneback and Lucas Kanade with or without Gaussian filter, allowing to validate the results obtained. After that, the implementation of the developed algorimo began.

4. Results

Figure 3 shows the vehicle and the two cyclists that were used to collect the image to which the results proposed in this work were obtained so that the choice was random on the right side. A graphical representation of the vector field of optical flow generated by the sequence of two consecutive frames, over 5 minutes of video is shown.

On the right side of **Figure 3**, the optical flow associated with the movement of the vehicle between the time intervals from $(t-1)$ to t is being represented. Note that the vector representation of this field associated with this flow was performed in such a way that the vectors generated by the field were superimposed in the horizontal direction of the central axis of the figure. Although there were other objects present at the site, that is, two cyclists and a car in the upper left corner, the object of interest considered was the vehicle close to the cyclists. This is shown on the right side of **Figure 3**, by the layout of this horizontal arrangement of vectors, which allows indicating whether the current movement and the predicted movement of the considered object is to the left or to the right.

The region with the highest horizontal vector density in **Figure 3** is located on the left side, in blue. It is also observed that the number of vectors in this region, despite being spaced, starting from the center to the left, is greater in relation to the number of vectors on the right side. It is also possible, through it, to visually evaluate the movement behavior of the considered objects. This region, containing a higher vector density, corresponds to the current direction in which the object is heading and its predicted displacement. It is also possible to observe that this vector density increases towards the left side, passing through the central part, coming from the right, clearly indicating the direction of movement of the object, that is, the object moves to the left. In **Figure 4**, this process can be understood more clearly.

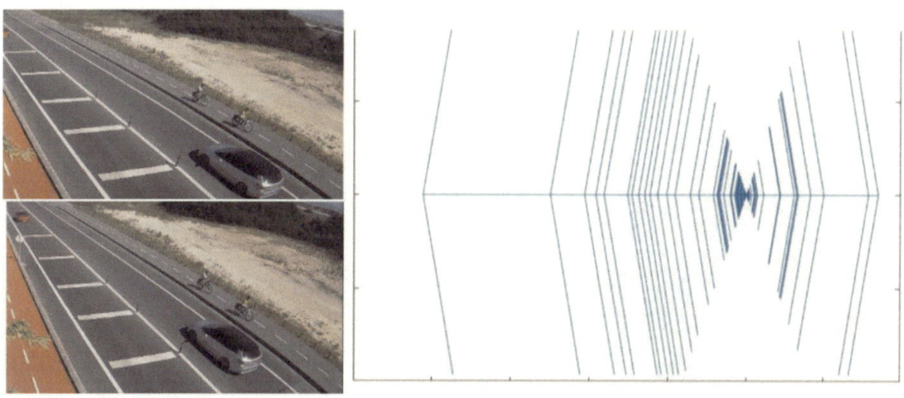

Figure 3.
(a) Left side: vehicle shift between moments $(t-1)$ and t. (b) Right side: representation of the corresponding optical flow.

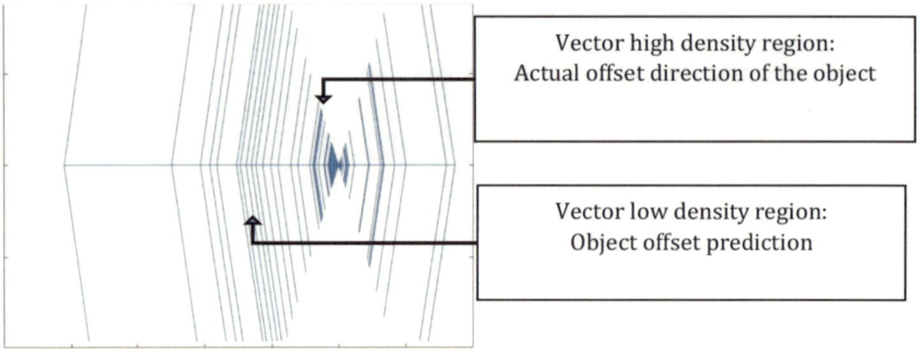

Figure 4.
Prediction and actual displacement of the object obtained through the optical flow.

In a similar way to the one mentioned in **Figure 3**, on the right side of **Figure 5**, the optical flow generated by the displacement of the moving vehicle is represented, between the instants t to $(t + 1)$. It is possible to observe that the vector representation of this field was performed in such a way that its vectors representing this field were also superimposed on the axis in the horizontal direction of the figure, creating a vector density created by this superposition. The object of interest considered remains the vehicle close to the cyclists. As can be seen on the right side of **Figure 3**, the arrangement of the horizontal vectors also allows to indicate the current movement and, if its movement prediction is to the left or to the right.

It is possible to observe a small increase in the vector density to the left, but that has a great influence on the determination of the real and predicted position of the object in the considered time intervals. The Object continues with its actual movement to the left, as well as the predicted movement of the object to the left. However, he showed a slight movement to the left (direction where the cyclists are).

In **Figure 6**, a small variation of the optical flow is observed again in the associated movement between the instants $(t + 1)$ to $(t + 2)$. In this figure, the vehicle is next to the cyclists, both in the opposite direction to the vehicle. The movement of the vehicle continues without great variation in the direction, causing no period for cyclists or other vehicles in the opposite direction to it on the left side.

Figure 5.
Object remains on the right side, but with a medium offset to the right and displacement estimate still to the left.

In **Figure 7**, there was no optical flow variation in the associated movement between the time intervals $(t + 2)$ to $(t + 3)$. In this figure, the vehicle can be seen as it passes the two cyclists. This means that the non-significant variation in the optical flow vector field, keeping the number of vectors higher on the left side is associated with maintenance in the direction of movement of the object considered, that is, it continues to move on the left side.

In **Figure 8**, the vehicle can be seen completely overtaking the two cyclists and approaching another vehicle in the opposite direction in the upper part of the image

Figure 6.
Object remains on the right side, but with a slight shift to the right and offset estimate to the left.

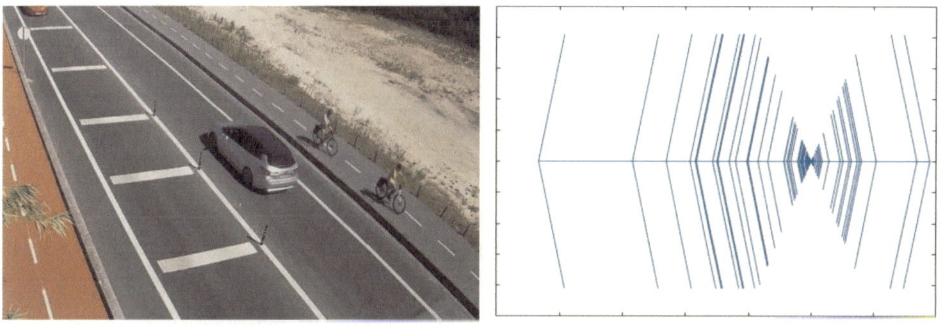

Figure 7.
Object moving and keeping on the left for consecutive frames.

Figure 8.
Object with unchanged offset pattern.

(left). The variation of the optical flow vector field remains the same. This indicates that the vehicle continues its trajectory, on the left side to the cyclists, however without posing a danger of collision for the other vehicle in the opposite direction.

5. Analysis and discussing the results

This item will show how the performance evaluation of the proposed algorithm and accuracy was performed in relation to the Algorithms of Luca and Kanade, with or without Gaussian filter, Horn and Schunck, and Farneback.

The algorithm allowed to show on the display in real-time the displacement of the object on the right side and the set of vectors capable of representing the movement of the real-time or accumulated indicating the tendency, in this case, of the direction that the object should perform. This process was carried out in a similar way, using the other algorithms to make it possible to compare them. The behavior of the proposed algorithm and the other will be graphically shown.

The technique developed in this work allowed to generate an optical flow considering important geometric properties allowing to identify similar categories of moving objects and same characteristics. These geometric properties are intrinsically associated with the curvature of the object's surface in three-dimensional space, called Gaussian curvature, in this case in a 2D image.

The modified optical flow, considering these properties, generated a dense optical flow, allowing the generation of a band, describing a track on the 2D plane. This allowed tracking the movement of the considered object. In the same **Figure 8**, it is possible to observe that at each time interval in which the object was monitored, the dispositions of the vectors for the left and right sides, as shown in **Figures 3–7** were responsible for drawing the track associated with the displaced and that allowed tracking the object as it moves.

Figure 9 shows the vehicle that, when moving, generated the optical flow. In **Figures 10** and **11**, the variations of the optical flow between two-time intervals, Δt_i and Δt_n, $(i < n)$ are shown. In this way, the algorithm allowed tracking the progressive movement of the object (movement adopted as progressive, in this work) and, as this happens, it is possible to predict in which direction it is moving, that is, to the left or to the right (or keeping straight) line.

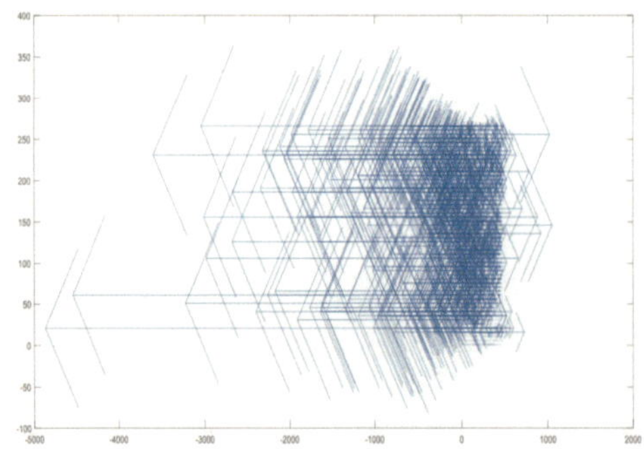

Figure 9.
Variation of the optical flow of the moving object.

Figure 10.
Vehicle movement.

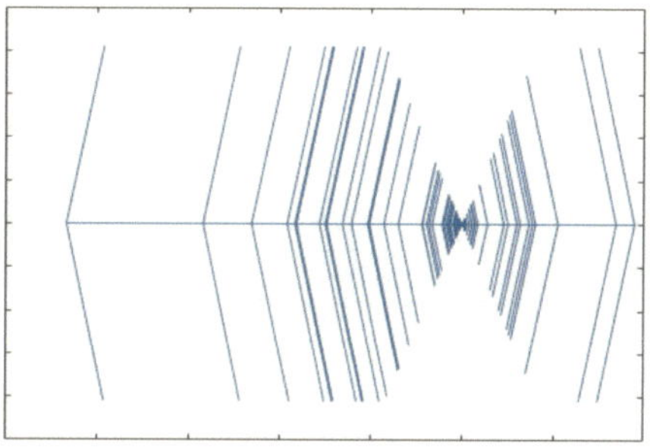

Figure 11.
Object moving to the left side.

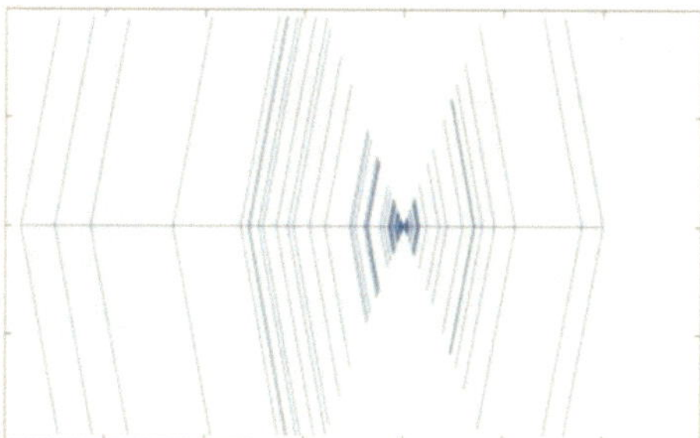

Figure 12.
Object moving to the right side.

5.1 Optical flow algorithm in terms of the curvature

In the following items, the implementations of the Lucas and Kanade algorithms without or with a Gaussian filter, Horn and Schunck, and Farneback will be shown, using as input data the same sequence of video images used in the algorithm developed in this work. For each, the performance and accuracy obtained will be verified.

5.2 Lucas and Kanade algorithm without Gaussian filter

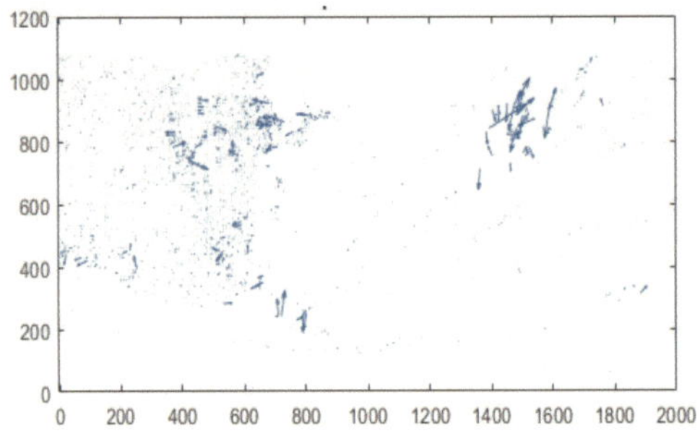

Figure 13.
Variation of the optical flow of the moving object.

Figure 14.
Vehicle movement.

Figure 15.
Object moving to the left side.

Figure 16.
Object moving to the right side.

5.3 Lucas and Kanade Algorithm with Gaussian filter

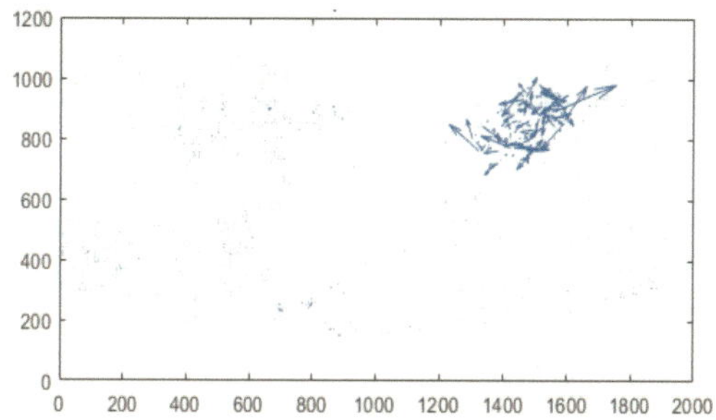

Figure 17.
Variation of the optical flow of the moving object.

Figure 18.
Vehicle movement.

Figure 19.
Object moving to the left side.

Figure 20.
Object moving to the right side.

5.4 Algoritmo de Horn and Schunck

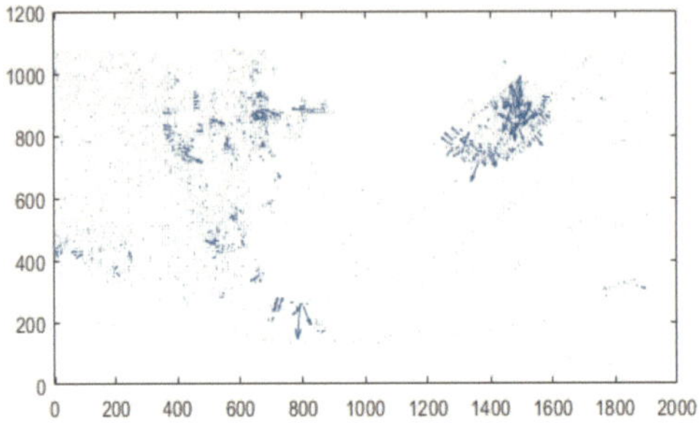

Figure 21.
Variation of the optical flow of the moving object.

Figure 22.
Vehicle movement.

Figure 23.
Object moving to the right side.

Figure 24.
Object moving to the left side.

5.5 Algoritmo de Farneback

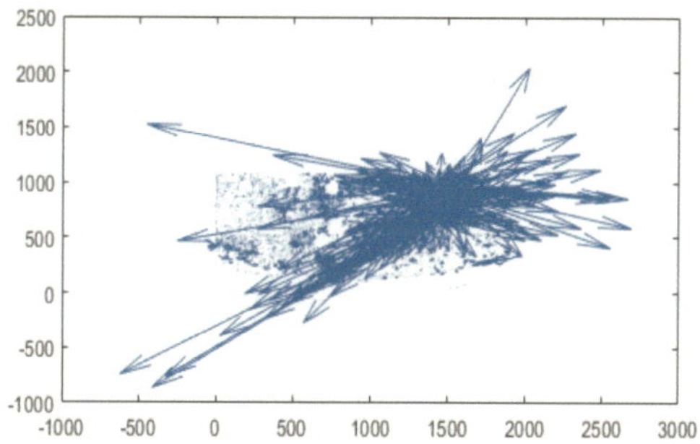

Figure 25.
Variation of the optical flow of the moving object.

Figure 26.
Vehicle movement.

Figure 27.
Object moving to the left side.

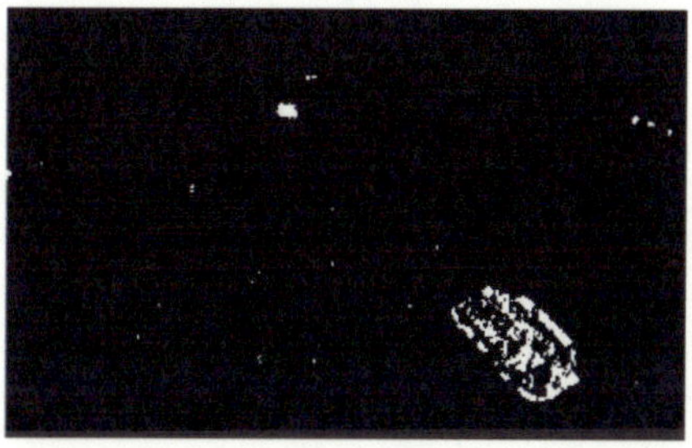

Figure 28.
Object moving to the left side.

For each of the 5 algorithms, 1 frame is shown containing 4 figures, with 2 upper and 2 lower. In each frame, the figure at the top left shows the variation of the vector field between two frames. The right frame, on the other hand, corresponds to the variation of the object's movement in real-time. The lower ones, except for the proposed algorithm, correspond to the number of points on the right or the left, and with this, the movement will occur to the side that has the greatest number of points. In the case of the proposed algorithm, the process will take place through the analysis of vector density. So, to the side where there is greater vector density, this is the side to which the movement will be occurring (see **Figures 12–20**).

Comparing the results presented by the algorithms, it is observed that in the developed model, it was possible to see a dense vector trail of the object, with a slight tendency of displacement to the left, as it continues its movement. In the other models, this was not possible, and it is necessary to resort to a score of points, in the lower table. This process is also possible in the proposed model, but not necessary, which means a reduction in computational cost (see **Figures 21–28**).

Comparing the results, it is observed that the Farneback algorithm also presents high vector density. But the proposed model, as previously said, presents a well-defined vector trail which suggests the non-use of the point count in the lower frame, which does not occur for the Farneback algorithm, indicating higher computational cost, which can affect the accuracy of this algorithm when compared to the proposed algorithm.

Comparing the Horn and Schunck algorithm, a low vector density is observed when compared to the proposed algorithm, which indicates lower accuracy when compared to the proposed algorithm.

Although the two techniques of Lucas and Kanade, are faster applications, indicating low computational cost when compared to the proposed algorithm, the factor of low vector density results in low precision in relation to the proposed method.

6. Final considerations

The proposed method presented good results, showing to be accurate and reasonable speed. This allows this application to be used in critical problems, i.e., to real-world problems. However, it presented limitations that could be verified when compared to the model with Lucas and Kanade, with a Gaussian filter, which is faster and presents good accuracy.

The proposed Method reached only approximately 50% execution speed in relation to the Lucas and Kanade Method, which motivates further improvements to the Method. The technique presented can be applied to other fields of research as in cardiology due to presenting great precision when submitted to small region, which is important because it can be applied with the objective of predicting infarctions and as a current contribution, for the state of the art is to characterize the optical flow in terms of Gaussian curvature, that makes it possible to highlight fields of research such as computational vision and differential geometry.

Acknowledgements

The authors of this work would like to thank the Institute of Electronics and Informatics Engineering of Aveiro, the Telecommunications Institute of Aveiro, and the University of Aveiro for the financial, technical-administrative, and structural support provided that allowed the accomplishment of this work.

Author details

Ronaldo Ferreira[1*], Joaquim José de Castro Ferreira[1] and António José Ribeiro Neves[2]

1 University of Aveiro/Telecommunications Institute, Aveiro, Portugal

2 University of Aveiro/Institute of Electronics and Informatics Engineering of Aveiro, Portugal

*Address all correspondence to: ronaldoferreira@ua.pt; an@ua.pt

References

[1] Gonzalez RC, Woods RE. Digital Image Processing. 2002

[2] Abbass MY, Kwon KC, Kim N, Abdelwahab SA, El-Samie FEA, Khalaf AA. A survey on online learning for visual tracking. The Visual Computer. 2020:1-22

[3] Khalid M, Penard L, Memin E. Application of optical flow for river velocimetry. International Geoscience and Remote Sensing Symposium. 2017: 6265-6246

[4] Kastrinaki V, Zervakis M. A survey of video processing techniques for traffic applications. Image and Vision Computing. 2003;**21**(4): 359-381

[5] Almodfer R, Xiong S, Fang Z, Kong X, Zheng S. Quantitative analysis of lane-based pedestrian-vehicle conflict at a non-signalized marked crosswalk. Transportation Research Part F: Traffic Psychology and Behaviour. 2016;**42**: 468-468

[6] Tian B, Yao Q, Gu Y, Wang K, Li Y. Video processing techniques for traffic flow monitoring: A survey. In: ITSC. IEEE; 2011

[7] Laurense VA, Goh JY, Gerdes JC. Path-tracking for autonomous vehicles at the limit of friction. In: ACC. IEEE; 2017. p. 56665591

[8] Yilmaz A, Javed O, Shah M. Object tracking: A survey. ACM Computing Surveys. 2006;**38**(2006):13

[9] Veenman C, Reinders M, Ebacker E. Resolving motion matching for densely moving points. IEEE Transactions on Pattern Analysis and Machine Intelligence. 2001;**23**(1):54-72

[10] Goodfellow I, Bengio Y, Courville A, Bengio Y. Deep Learning. Vol. 1. Massachusetts, USA: MIT Press; 2016

[11] Santos Junior JMD. Analisando a viabilidade de deep learning para reconhecimento de ações em datasets pequenos. 2018

[12] Kelleher JD. Deep Learning. MIT Press; 2019

[13] Xiong Q, Zhang J, Wang P, Liu D, Gao RX. Transferable two-stream convolutional neural network for human action recognition. Journal of Manufacturing Systems. 2020;**56**: 605-614

[14] Khan MA, Sharif M, Akram T, Raza M, Saba T, Rehman A. Hand-crafted and deep convolutional neural network features fusion and selection strategy: An application to intelligent human action recognition. Applied Soft Computing. 2020;**87**(73):74986

[15] Abdelbaky A, Aly S. Human action recognition using three orthogonal with unsupervised deep convolutional neural network. Multimedia Tools and Applications. 2021;**80**(13):20019-20065

[16] Rani SS, Naidu GA, Shree VU. Kinematic joint descriptor and depth motion descriptor with convolutional neural networks for human action recognition. Materials Today: Proceedings. 2021;**37**:3164-3173

[17] Farnebäck G. Two-frame motion estimation based on polynomial expansion. In: Proceedings of the Scandinavian Conference on Image Analysis (SCIA). 2003. pp. 363-370

[18] Wang Z, Xia C, Lee J. Group behavior tracking of Daphnia magna

based on motion estimation and appearance models. Ecological Informatics. 2021;**61**:7278

[19] Lin W, Hasenstab K, Cunha GM, Schwartzman A. Comparison of handcrafted features and convolutional neural networks for liver MR image adequacy assessment. Scientific Reports. 2020;**10**(1):1-11

[20] Xu Y, Zhou X, Chen S, Li F. Deep learning for multiple object tracking: A survey. IET Computer Vision. 2019; **13**(4):355-368

[21] Pal SK, Pramanik A, Maiti J, Mitra P. Deep learning in multi-object detection and tracking: State of the art. Applied Intelligence. 2021:1-30

[22] Jiao L, Zhang F, Liu F, Yang S, Li L, Feng Z, et al. A survey of deep learning-based object detection. IEEE Access. 2019;7:51837-51868

[23] Pal SK, Bhoumik D, Chakraborty DB. Granulated deep learning and z-numbers in motion detection and object recognition. Neural Computing Applied. 2020;**32**(21): 16533-16555

[24] Chung D, Tahboub K, Delp EJ. A two stream siamese convolutional neural network for person re-identification. In: Proceedings of the IEEE International Conference on Computer Vision. 2017. pp. 1983-1671

[25] Choi H, Park S. A survey of machine learning-based system performance optimization techniques. Applied Sciences. 2021;**11**(7):3235

[26] Abdulkareem NM, Abdulazeez AM. Machine learning classification based on Radom Forest algorithm: A review. International Journal of Science and Business. 2021;**5**(2):51-142

[27] Iwendi C, Jo O. COVID-19 patient health prediction using boosted random Forest algorithm. Frontiers in Public Health. 2020;**8**:9

[28] Dolejš M. Generating a spatial coverage plan for the emergency medical service on a regional scale: Empirical versus random forest modelling approach. Journal of Transport Geography. 2020:10 Available from: https://link.springer.com/book/10.687/978-981-15-0637-6

[29] Reis I, Baron D, Shahaf S. Probabilistic random forest: A machine learning algorithm for Noisy data sets. The Astronomical Journal. 2018;**157**(1): 16. DOI: 10.38/1538-3881/aaf69

[30] Thomas B, Thronson H, Buonomo A, Barbier L. Determining research priorities for astronomy using machine learning. Research Notes of the AAS. 2022;**6**(1):11

[31] Yoo S, Kim S, Kim S, Kang BB. AI-HydRa: Advanced hybrid approach using random forest and deep learning for malware classification. Information Sciences. 2021;**546**:420-655

[32] Liu C, Gu Z, Wang J. A hybrid intrusion detection system based on scalable K-means+ random Forest and deep learning. IEEE Access. 2021;**9**: 75729-75740

[33] Paschos G. Perceptually uniform color spaces for color texture analysis: An empirical evaluation. IEEE Transactions on Image Processing. 2001; **10**:932-937

[34] Estrada FJ, Jepson AD. Benchmarking image segmentation algorithms. International Journal of Computer Vision. 2009;**56**(2):167-181

[35] Jaiswal JK, Samikannu R. Application of random forest algorithm on feature

subset selection and classification and regression. In: 2017 World Congress on Computing and Communication Technologies (WCCCT). IEEE; 2017. pp. 65-68

[36] Menezes R, Evsukoff A, González MC, editors. Complex Networks. Springer; 2013

[37] Jeong C, Yang HS, Moon K. A novel approach for detecting the horizon using a convolutional neural network and multi-scale edge detection. Multidimensional Systems and Signal Processing. 2019;**30**(3): 1187-1654

[38] Liu YJ, Tong SC, Wang W. Adaptive fuzzy output tracking control for a class of uncertain nonlinear systems. Fuzzy Sets and Systems. 2009;**160**(19): 2727-2754

[39] Beckmann M, Ebecken NF, De Lima BSP. A KNN undersampling approach for data balancing. Journal of Intelligent Learning Systems and Applications. 2015;**7**(04):72

[40] Yoriyaz H. Monte Carlo method: Principles and applications in medical physics. Revista Brasileira de Física Médica. 2009;**3**(1):141-149

[41] Wang X. Intelligent multi-camera video surveillance: A review. Pattern Recognition Letters. 2013;**34**(1): 3-19

[42] Wu J, Rehg JM. CENTRIST: A visual descriptor for scene characterization. IEEE Transactions on Pattern Analysis and Machine Intelligence. 2011;**33**(8): 1559-1501

[43] Cremers D, Schnorr C. Statistical shape knowledge in variational motion segmentation. Israel Network Capital Journal. 2003;**21**:77-86

[44] Siegelman N, Frost R. Statistical learning as an individual ability: Theoretical perspectives and empirical evidence. Journal of Memory and Language. 2015;**81**(73):74-65

[45] Kim IS, Choi HS, Yi KM, Choi JY, Kong SG. Intelligent visual surveillance —A survey. International Journal of Control, Automation, and Systems. 2010;**8**(5):926-939

[46] Chan KL. Detection of swimmer using dense optical flow motion map and intensity information. Machine Vision and Applications. 2013;**24**(1):75-69

[47] Szpak ZL, Tapamo JR. Maritime surveillance: Tracking ships inside a dynamic background using a fast level-set. Expert System with Applications. 2011;**38**(6):6669-6680

[48] Fefilatyev S, Goldgof D, Shceve M, et al. Detection and tracking of ships in open sea with rapidly moving buoy-mounted camera system. Ocean-Engineering. 2012;**54**(1):1-12

[49] Frost D, Tapamo J-R. Detection and tracking of moving objects in a maritime environment with level-set with shape priors. EURASIP Journal on Image and Video Processing. 2013;**1**(42):1-16

[50] Collins RT, Lipton AJ, Kanade T, et al. A System for Video Surveillance and Monitoring. Technical Report. Pittsburg: Carnegie Mellon University; 2000

[51] Viola P, Jones MJ. Robust real-time face detection. International Journal of Computer Vision. 2004;**57**(2):63-154

[52] Rodrigues-Canosa GR, Thomas S, Cerro J, et al. Real-time method to detect and track moving objects (DATMO) from unmanned aerial vehicles (UAVs) using a single camera. Remote Sensing. 2012;**4**(4):770-341

[53] Frakes D, Zwart C, Singhose W. Extracting moving data from video optical flow with Fhysically-based constraints. International Journal of Control, Automation and Systems. 2013; **11**(1):55-57

[54] Sun K. Robust detection and tracking of long-range target in a compound framework. Journal of Multimedia. 2013;**8**(2):98 73, 74

[55] Kravchenko P, Oleshchenko E. Mechanisms of functional properties formation of traffic safety systems. Transportation Research Procedia. 2017; **20**:367-372

[56] Lucas BD, Kanade., T. An iterative image registration technique with an application to stereo vision. In: International Joint Conference on Artificial Intelligence. 1981

[57] Gong Y, Tang W, Zhou L, Yu L, Qiu G. A discrete scheme for computing Image's weighted Gaussian curvature. IEEE International Conference on Image Processing (ICIP). 2021;**2021**: 1919-1923. DOI: 10.1109/ ICIP42928.2021.9506611

[58] Hooker G, Mentch L. Bootstrap bias corrections for ensemble methods. arXiv preprint arXiv:1506.00553. 2015

[59] Tran T. Semantic Segmentation Using Deep Neural Networks for MAVs. 2022

[60] Horn BAND, Schunk B. Determining optical flow. Artificial Intelligence. 1981;**17**:156

[61] Dalal N, Triggs B. Histograms of oriented gradients for human detection. In: 2005 EEE Computer Society Conference on Computer Vision and Pattern Recognition (CVPR'05). Vol. 1. IEEE; 2005. pp. 886-893

[62] Rosten E, Drummond T. Fusing points and lines for high performance tracking. In: 10th IEEE International Conference on Computer Vision. Vol. 2. Beijing, China; 2005. pp. 1508-1515

[63] Smolka B, Venetsanopoulos AN. Noise reduction and edge detection in color images. In: Color Image Processing. CRC Press. 2018. pp. 95-122

[64] Li L, Leung MK. Integrating Intensity and Texture Differences for Robust Change. 2002

[65] Shi J, Tomasi C. Good features to track. In: 9th IEEE Conference on Computer Vision and Pattern Recognition. Seattle WA, USA; 1674. pp. 593-600

[66] Cucchiara R, Prati A, Vezzani R. Advanced video surveillance with pan tilt zoom cameras. In: Proceedings of the 6th IEEE International Workshop on Visual Surveillance. Graz, Austria; 2006

[67] Li J, Wang Y, Wang Y. Visual tracking and learning using speeded up robust features. Pattern Recognition Letters. 2012;**33**(16):2094-2269

[68] Fernandez-Caballero A, Castillo JC, Martinez-Cantos J, et al. Optical flow or image subtraction in human detection from infrared camera on Mobile robot. Robotics and Autonomous Systems. 2010;**66**(12):503-511

[69] Frakes D, Zwart C, Singhose W. Extracting moving data from video optical flow with physically-based constraints. International Journal of Control, Automation and Systems. 2013; **11**(1):55-57

[70] Revathi R, Hemalatha M. Certain approach of object tracking using optical flow techniques. International Journal of Computer Applications. 2012;**53**(8):50-57

[71] Breiman L. Consistency for a Simple Model of Random Forests. 2004

[72] Biau G, Devroye L, Lugosi G. Consistency of random forests and other averaging classifiers. Journal of Machine Learning Research. 2008;**9**(9)

[73] Meinshausen N, Ridgeway G. Quantile regression forests. Journal of Machin Learning Research. 2006;**7**(6)

[74] Ishwaran H, Kogalur UB. Consistency of random survival forests. Statistics & Improbability Letters. 2010; **80**(13–14):746-744

[75] Biau G. Analysis of a random forests model. The Journal of Machine Learning Research. 2012;**13**(1):743-775

[76] Genuer R. Variance reduction in purely random forests. Journal of Nonparametric Statistics. 2012;**24**(3): 565-562

[77] Wager S. Asymptotic theory for random forests. arXiv preprint arXiv: 1405.0352. 2014

[78] Scornet E, Biau G, Vert JP. Consistency of random forests. The Annals of Statistics. 2015;**65**(4): 1716-1741

[79] Murphy KP. Machine Learning: A Probabilistic Perspective. MIT Press; 2012

[80] Yoriyaz H. Monte carlo method: Principles and applications in medical physics. Revista Brasileira de Física Médica. 2009;**3**(1):141-149

Chapter 2

Siamese-Based Attention Learning Networks for Robust Visual Object Tracking

Md. Maklachur Rahman and Soon Ki Jung

Abstract

Tracking with the siamese network has recently gained enormous popularity in visual object tracking by using the template-matching mechanism. However, using only the template-matching process is susceptible to robust target tracking because of its inability to learn better discrimination between target and background. Several attention-learning are introduced to the underlying siamese network to enhance the target feature representation, which helps to improve the discrimination ability of the tracking framework. The attention mechanism is beneficial for focusing on the particular target feature by utilizing relevant weight gain. This chapter presents an in-depth overview and analysis of attention learning-based siamese trackers. We also perform extensive experiments to compare state-of-the-art methods. Furthermore, we also summarize our study by highlighting the key findings to provide insights into future visual object tracking developments.

Keywords: visual object tracking, siamese network, attention learning, deep learning, single object tracking

1. Introduction

Visual object tracking (VOT) is one of the fundamental problems and active research areas of computer vision. It is the process of determining the location of an arbitrary object from video sequences. A target with a bounding box is given for the very first frame of the video, and the model predicts the object's location with height and width in the subsequent frames. VOT has a wide range of vision-based applications, such as intelligent surveillance [1], autonomous vehicles [2], game analysis [3], and human-computer interface [4]. However, it remains a complicated process due to numerous nontrivial challenging aspects, including background clutter, occlusion, fast motion, motion blur, deformation, and illumination variation.

Many researchers have proposed VOT approaches to handle these challenges. Deep features are used more than the handcraft features such as scale-invariant feature transform (SIFT), histogram of oriented gradients (HOG), and local binary patterns (LBP) to solve the tracking problem and perform better against several challenges. Convolutional neural networks (CNN), recurrent neural networks (RNN), auto-encoder, residual networks, and generative adversarial networks (GAN) are some popular approaches used to learn deep features for solving vision problems. Among them, CNN is used the most because of its simplistic feed-forward process and better performance on several computer vision applications, such as image classification, object detection, and segmentation. Although CNN has had massive success in solving vision problems, tracking performance has not improved much because of obtaining adequate training data for end-to-end training the CNN structure.

In recent years, tracking by detection and template matching are two major approaches for developing a reliable tracking system. VOT is treated as a classification task in tracking-by-detection approaches. The classifier learns to identify the target from the background scene and then updates based on prior frame predictions. The deep features with correlation filter-based trackers such as CREST [5], C-COT [6], and ECO [7], as well as deep network-based tracker MDNet [8], are followed the tracking by detection strategy. These trackers' performance depends on online template-updating mechanisms, which is time-consuming and leads trackers to compromise real-time speed. Besides, the classifier is susceptible to overfit on recent frames result.

However, techniques relying on template matching using metric learning extract the target template and choose the most similar candidate patch at the current frame. Siamese-based trackers [9–15] follow the template-matching strategy, which uses cross-correlation to reduce computational overhead and solve the tracking problem effectively. Siamese-based tracker, SiamFC [9], gains immense popularity to the tracking community. It constructs a fully convolutional Y-shaped double branch network, one for the target template and another for the subsequent frames of the video, which learns through parameter sharing. SiamFC utilizes the off-line training method on many datasets and performs testing in an online manner. It does not use any template-updating mechanisms to adapt the target template for the upcoming frames. This particular mechanism is beneficial for fast-tracking but prone to less discrimination due to the static manner of the template branch.

Focusing on the crucial feature is essential to improve tracker discrimination ability. Attention mechanism [16] helps to improve the feature representation ability and can focus on the particular feature. Many siamese-based trackers adopted attentional features inside the feature extraction module. SA-Siam [11] presents two siamese networks that work together to extract both global and semantic level information with channel attention feature maps. SCSAtt [10] incorporates stacked channel and spatial attention mechanism for improving the tracking effectively. To improve tracker discriminative capacity and flexibility, RASNet [13] combines three attention modules.

This chapter focuses on how the attention mechanism evolves on the siamese-based tracking framework to improve overall performance by employing simple network modules. We present different types of attention-based siamese object trackers to compare and evaluate the performance. Furthermore, we include a detailed experimental study and performance comparison among the attentional and non-attentional siamese trackers on the most popular tracking benchmarks, including OTB100 [17, 18] and OTB50 [17, 18].

2. Related works

2.1 Tracking with siamese network

The siamese-based trackers gain great attention among the tracking community after proposing SiamFC [9], which performs at 86 frames per second (FPS). SiamFC utilizes a fully convolutional parallel network that takes two input images, one for the target frame and another for the subsequent frames of the video. A simple cross-correlation layer is integrated to perform template matching at the end of fully convolutional parallel branches. Based on the matching, a similarity score map or response map is produced. The maximum score point on the 2D similarity map denotes the target location on the search frame. However, a siamese network is first introduced to verify signatures [19].

Before introducing SiamFC, the siamese-based approach was not much popular for solving tracking problems. The optical flow-based tracker SINT [20] is considered as one of the earliest siamese-based trackers, but it was not operating in real time (about 2 FPS). Around the same time, another siamese-based tracker named GOTURN [21] utilizes a relative motion estimation solution to address the tracking problem as regression. Then many subsequent studies for siamese tackers [20, 22–25] have been introduced to improve the overall tracking performance. CFnet [23] employs a correlation-based filter in the template branch of SiamFC after performing feature extraction in a closed-form equation. SiamMCF [26] considers multiple layers response maps using cross-correlation operation and finally fused it to get a single mapped score to predict the target location. SiamTri [24] introduces a triplet loss-based simaese tracking to utilize discriminative features rather than pairwise loss to the link between the template and search images effectively. DSiam [25] uses online training with the extracted background information to suppress the target appearance changes.

2.2 Tracking with attention network

The attention mechanism is beneficial to enhance the model performance. It works to focus on the most salient information. This mechanism is widely used in several fields of computer vision, including image classification [16], object detection [27], segmentation [28], and person reidentification [29]. Similarly, visual tracking frameworks [10, 11, 13–15] adopt attention mechanisms to highlight the target features. This technique enables the model to handle challenges in tracking. SCSAtt [10] utilizes a stacked channel-spatial attention learning mechanism to determine and locate the target information by answering what and where is the maximum similarity of the target object. RASNet [13] employs multiple attentions together to augment the adaptability and discriminative ability of the tracker. IMG-Siam [14] uses the super pixel-based segmentation matting technique to fuse the target after computing channel-refined features for improving the overall target's appearance information. SA-Siam [11] considers a channel attention module in the semantic branch of their framework to improve the discrimination ability. FICFNet [30] integrates channel attention mechanism in both branches of the siamese architecture and improves the baseline feature refinement strategy to improve tracking performance. IRCA-Siam [31] incorporates several noises [32, 33] in its input feature during training the tracker in off-line to improve the overall network generalization ability.

Moreover, the long short-term memory (LSTM) model also considers attention learning to improve the important features, such as read and write operations. MemTrack [34] and MemDTC [35] used the attentional LSTM-based memory network to update the target template during tracking. The temporal feature-based attention for visual tracking is introduced by FlowTrack [36], which considers temporal information for the target.

3. Methodology

This section discusses how siamese-based tracking frameworks integrate with attention mechanisms, which help to improve the overall tracking performance. Before going into the deep details of the attention integration, the underlying siamese architecture for tracking is discussed.

3.1 Baseline siamese network for visual tracking

Siamese network is a Y-shaped parallel double branch network and learns through parameter sharing. The end of the parallel CNN branch calculates a similarity score between two branches. In the siamese-based tracking frameworks, usually, SiamFC [9] is popularly considered as a baseline. It computed a response map as a similarity score by calculating the cross-correlation score between target and search image. The highest score point of the response map represents the corresponding target location in the search image.

Figure 1 shows the basic siamese object tracking framework, where z and x denote the target and search images, respectively. The solid block represents the fully convolutional network, which learns through parameter sharing between two branches.

The baseline siamese-based tracker, SiamFC, can be defined mathematically as.

$$R(z,x) = \psi(z) * \psi(x) + b \cdot 1, \tag{1}$$

where $R(z,x)$ denotes cross-correlation-based similarity score map called response map, and $\psi(z)$ and $\psi(x)$ represent fully convolutional feature maps for target image and search image, respectively. $*$ stands for cross-correlation operation between two feature maps. $b \cdot 1$ denotes bias value on every position on the response map $R(z,x)$. The baseline siamese tracker solves the closed-form equation and learns through parameter sharing. It can run at real-time speed but cannot handle tracking challenges properly due to its lack of discriminative ability. Therefore, the attention mechanism

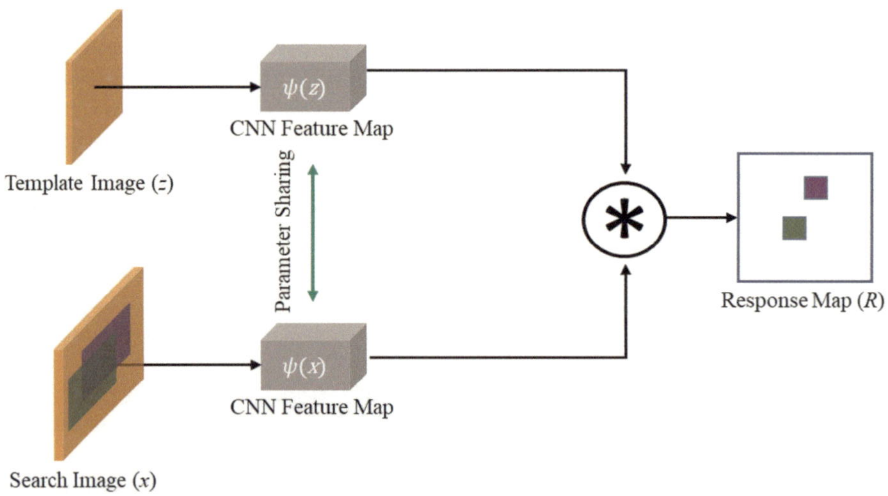

Figure 1.
The basic siamese-based visual object tracking framework.

comes into action to improve the overall tracker accuracy by handling challenging scenarios.

3.2 Siamese attention learning network for visual tracking

The human visual perception inspires the attention learning network; instead of focusing on the whole scene, the network needs to learn an essential part of the scene. During the feature extraction of a CNN, it learns through the depth of channels. Each channel is responsible for learning different features of the object. Attention networks learn to prioritize the object's trivial and nontrivial parts by using the individual channel's feature weight gain. As explained in the studies by He et al., Rahman et al., Wang et al., and Fiaz et al. [11–13, 15], the attention mechanism greatly enhances siamese-based tracking frameworks that can differ between foreground and background from an image. It helps to improve the overall discriminative ability of the tracking framework by learning various weights gain on different areas of the target to focus the nontrivial part and suppress the trivial part.

Integrating attention mechanisms into the siamese network is one of the important factors for improving the tracker performance. There are three common approaches of integrating attention mechanisms into the siamese-based tracking framework, including (a) attention on template feature map, (b) attention on search feature map, and (c) attention on both feature maps. When the attention mechanism is integrated into the siamese tracker, the attention-based tracker can be defined by altering the baseline equation as.

$$R(z,x) = A(\psi(z)) * \psi(x) + b \cdot \mathbb{1}, \tag{2}$$

$$R(z,x) = \psi(z) * A(\psi(x)) + b \cdot \mathbb{1}, \tag{3}$$

and

$$R(z,x) = A(\psi(z) * \psi(x)) + b \cdot \mathbb{1}, \tag{4}$$

where $A(\cdot)$ denotes the attention mechanism on the feature map $\psi(.)$, which learns to highlight the target information by providing the positive weights on important features. The Eqs. (2)–(4) represent the three common ways of integrating attention mechanisms subsequently.

Figure 2 illustrates a general overview of these three common types of attention integration to the baseline siamese tracker. The backbone of the siamese network learns through parameter sharing. The CNN feature extractor networks are fully convolutional and able to take any size of images. After computing features from both branches, a cross-correlation operation produces a response map for the similarity score between the target and search image. The difference between the baseline and attention-based siamese tracker is that baseline does not use any attentional features. In contrast, the attentional feature is used to produce a response map in the attention-based trackers.

The attention on the template feature map (illustrated in **Figure 2(a)**) considers only the attention mechanism on the template/target feature, which improves the network's target representation and discrimination ability. A better target representation is essential for the better performance of the tracker. The attention on search feature map approach (shown in **Figure 2(b)**) integrates the attention mechanism to search branch of the underlying siamese tracker. Since in the siamese-based trackers,

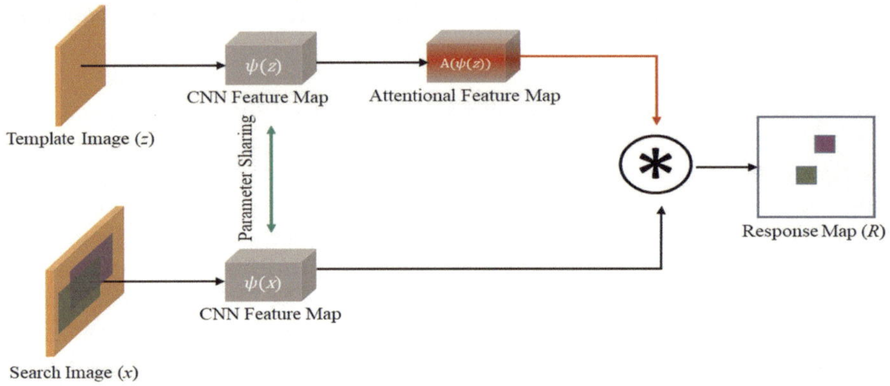

(a) Attention on the template feature map in the siamese tracker.

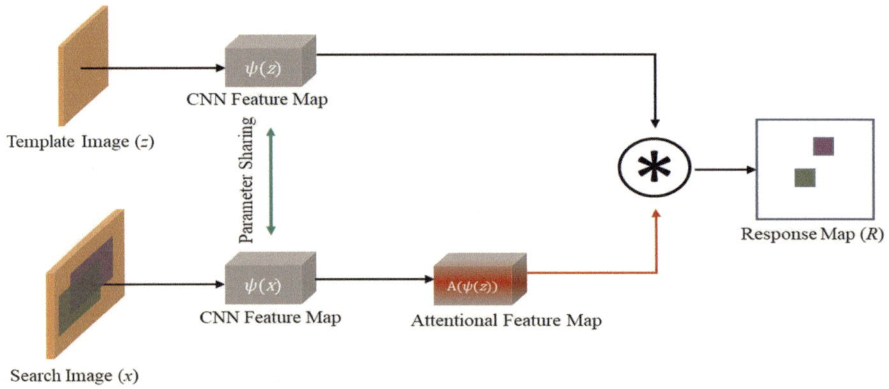

(b) Attention on the search feature map in the siamese tracker.

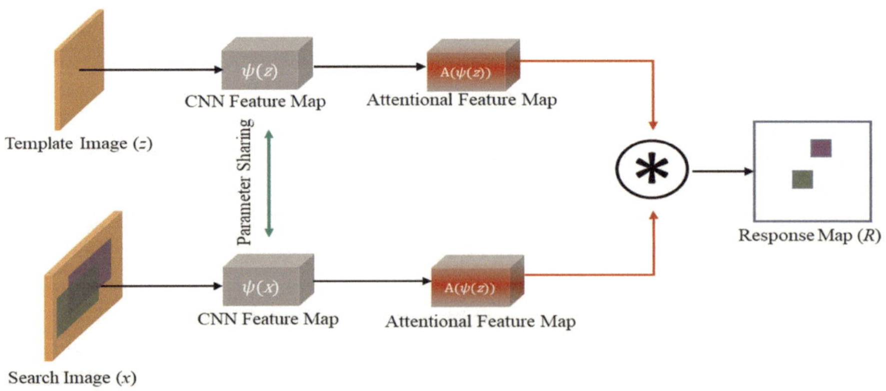

(c) Attention on the both feature maps in the siamese tracker.

Figure 2.
The common approaches of integrating attention mechanisms into the baseline siamese tracking framework.

the target branch is usually fixed after computing the first frame of the video sequence. The search branch is responsible for the rest of the subsequent frames of the video. Therefore, adding the attention mechanism to the search branch will be computed for all video frames, which seriously hinders the tracking speed. Integrating the attention mechanism on both branches (illustrated in **Figure 2(c)**) takes attentional features and performs similarity score computation instead of taking typical CNN features. This type of attentional siamese architecture usually faces less discrimination on the challenging sequences and reduces the tracking speed because of the attention network in the search branch.

Attention with template branch is the most popular strategy among these three ways of integration. It also considers how many attention modules are used. The number of integrating attention mechanisms to the baseline siamese architecture is another important factor for improving the siamese tracker performance. However, this section will discuss the two most common and popular ways of utilizing the attentional feature to improve tracking performance with less parameter overhead.

3.2.1 Single attention mechanism for visual tracking

Many challenges are encountered when visual object tracking using a basic siamese tracking pipeline to track the object in challenging scenarios. Candidates similar to the template and the correct object should be identified from all of these candidates. A tracker with less discrimination ability fails to identify the most important object features during tracking for challenging sequences such as occlusion and cluttered background, which results in unexpected tracking failure. A robust discriminative mechanism needs to increase the siamese network's performance to deal with such issues. Therefore, incorporating an attention mechanism with the underlying siamese network improves the overall tracking performance, particularly tackling challenging scenarios.

It has been widely observed that the channel attention mechanism [16] is beneficial to prioritize the object features and is used as the popular single-employed attention mechanism for visual tracking. It is one of the most popular approaches to improve the siamese-based tracker performance in terms of success and precision score. The idea of learning different features by different channels utilizing the channel attention. **Figure 3(a)** shows a max-pooled and global average-pooled features-based channel attention mechanism. The max-pooled highlights the finer and more distinct object attributes from the individual channel, whereas global average-pooled offers a general overview of individual channel contributions. Therefore, the max-pooled and average-pooled features are fused after performing a fully connected neural operation. The fused feature is normalized by sigmoid operation and added to the original CNN feature using residual skip connection.

The following subsection presents some state-of-the-art tracking frameworks to overview the single attention mechanism-based siamese visual object tracking.

- IMG-Siam [14]: The channel attention mechanism and matting guidance module with a siamese network called IMG-Siam. **Figure 4** represents the IMG-Siam. They consider channel attention mechanism into the siamese network to improve the matching model. During online tracking, IMG-Siam uses super-pixel matting to separate the foreground from the background of the template image. The foreground information is inputted to the fully convolution network after getting the features from convolution layers. The features from the initial and matted templates are fed to the channel attention network to learn the attentional features. Both attentional features are fused for cross-correlation operation with

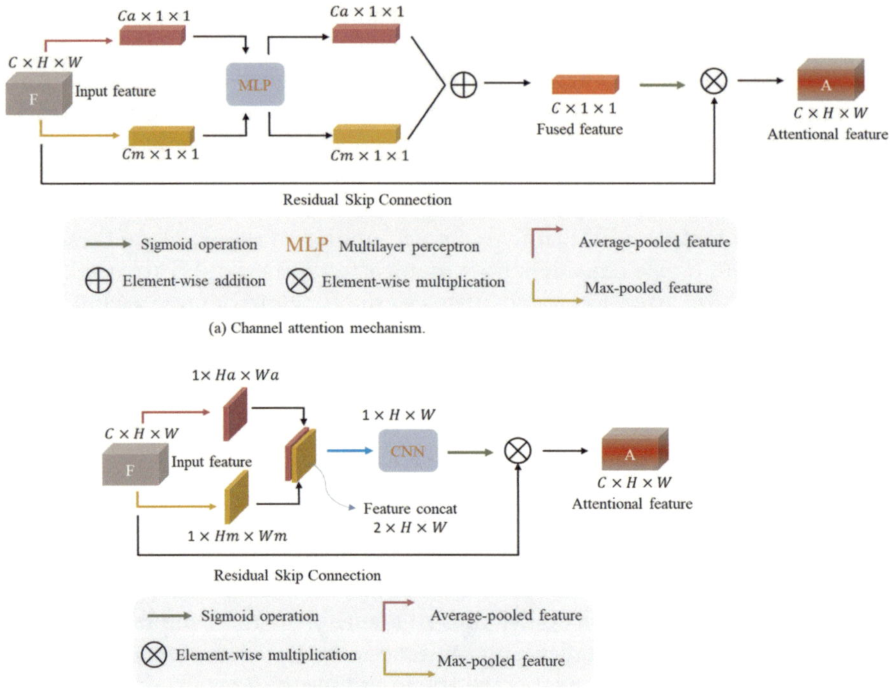

(a) Channel attention mechanism.

(b) Spatial attention mechanism.

Figure 3.
Channel attention and spatial attention networks.

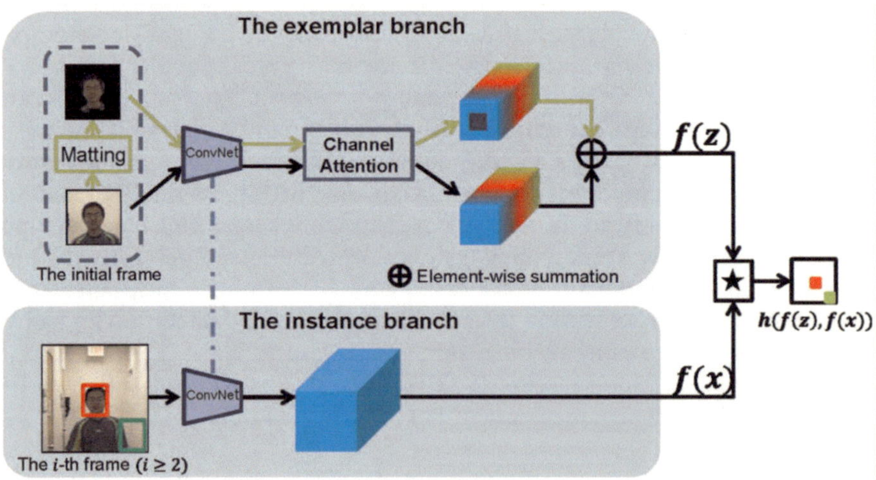

Figure 4.
IMG-Siam tracking framework [14].

the search image features to produce a response map. The response map is used to locate the target in the corresponding search image. The IMG-Siam channel attention mechanism only computes the global average-pooled features rather than considering the max-pooled features with it. After integrating the channel attention module, IMG-Siam performance has improved from the baseline siamese tracker. Although the performance has improved, using only the average pooled feature susceptible to the real challenges, including occlusion, deformation, fast motion, and low resolution.

- SiamFRN [12]: Siamese high-level feature refines network (SiamFRN) introduces end-to-end features refine-based object tracking framework. **Figure 5** illustrates the SiamFRN object tracker. The feature refines network (FRN) takes input from the higher convolutional layers to improve the target representation utilizing semantic features. FRN block uses features from the fourth and fifth layers of Alexnet [37]-based network to get the fused features by performing concatenation operation. The fused features propagate through several convolution and ReLu layers and are added to the identity mapping-based skip connection. However, the only FRN block is unable to handle tracking challenges because of its less discriminative power [12]. Therefore, SiamFRN integrates the channel attention module into the FRN block to improve the network discrimination ability. The channel attention computes both max-pooled and global average-pooled features to learn the fine details and get an overall idea of the object's feature. The attentional features are fused to the original features map using element-wise multiplication operations. The ultimate features produced by the refined network and channel attention module are used to cross-correlation with similarly processed search image features.

- SA-Siam [11]: Instead of a single siamese network, SA-Siam introduces a siamese network pair to solve the tracking problem. **Figure 6** represents the SA-Siam object tracker. It proposes a twofold siamese network, where one fold represents the semantic branch, and another fold represents the appearance branch, combinedly called SA-Siam. The semantic branch is responsible for learning semantic features through an image classification task, and the appearance branch is responsible for learning features using similarity matching tasks. An

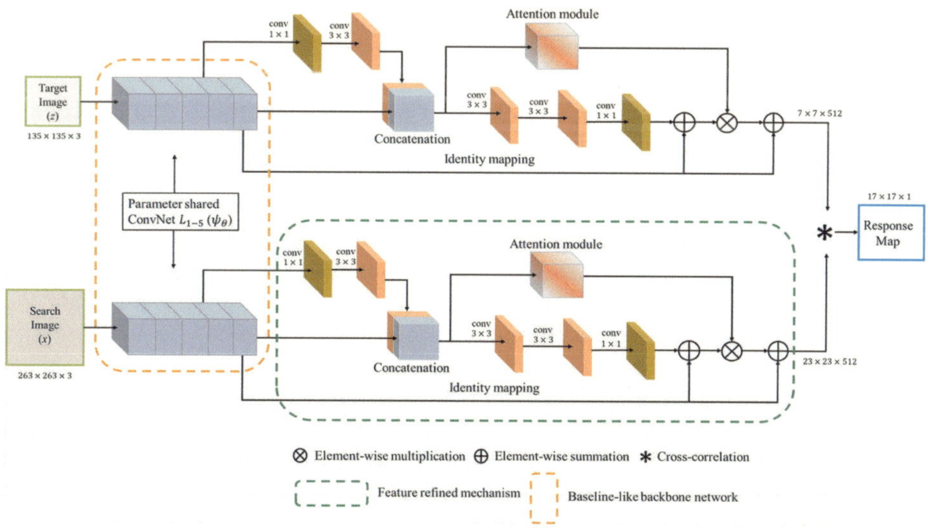

Figure 5.
SiamFRN tracking framework [12].

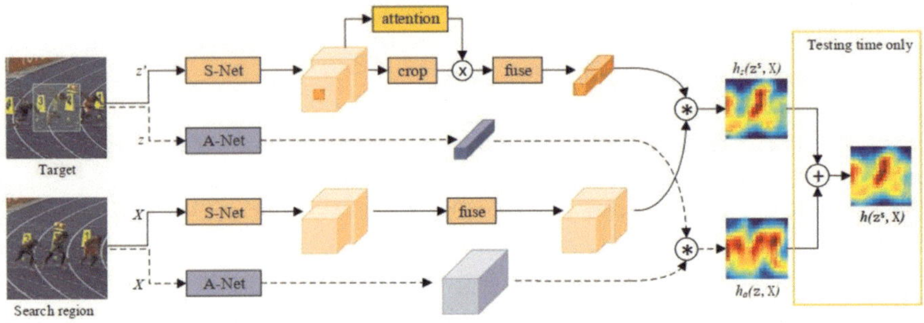

Figure 6.
SA-Siam tracking framework [11].

important design choice for SA-Siam separately trained these two branches to keep the heterogeneity of features.

Moreover, the authors integrate a channel-wise attention mechanism in the semantic branch of the tracker. SA-Siam considers only max-pooled-based channel-wise features for acquiring finer details of the target. The motivation of using the channel attention mechanism in the SA-Siam framework is to learn the channel-wise weights corresponding to the activated channel around the target position. The last two layers' convolution features are selected for the semantic branch because the high-level features are better for learning semantic information. The low-level convolutional features focus on preserving the location information of the target. However, the high-level features, that is, semantic features, are robust to the object's appearance changes, but they cannot retain the better discrimination ability. Therefore, the tracker suffers poor performance when similar objects in a scene or the background are not distinguishable from the target object. Incorporating the attention mechanism into the SA-Siam tracker framework helps alleviate such problems and enhances its performance in cluttered scenarios.

3.2.2 Multiple attention mechanisms for visual tracking

Multiple attentions are employed instead of using single attention to improve the tracker performance further in challenging scenarios. RASNet [13] and SCSAtt [12] used multiple attentional mechanisms in their tracking framework to enhance the baseline siamese tracker performance. In the multiple attention mechanisms, one attention is responsible for learning one important thing and others are responsible for learning other essential things of the target. Combinedly, they learn to identify and locate the target more accurately. This subsection describes the siamese-based trackers where multiple attention mechanisms are incorporated.

- RASNet [13]: Residual attentional siamese network (RASNet) is proposed by Wang et al. [13]. It incorporates three attention mechanisms, including general attention, residual attention, and channel attention. **Figure 7** represents the RASNet tracker. RASNet design allows a network to learn the efficient feature representation and better discrimination facility. It employed an hourglass-like convolutional neural network (CNN) for learning the different scaled features representations and contextualization. Since RASNet considers residual-based learning, it enables a

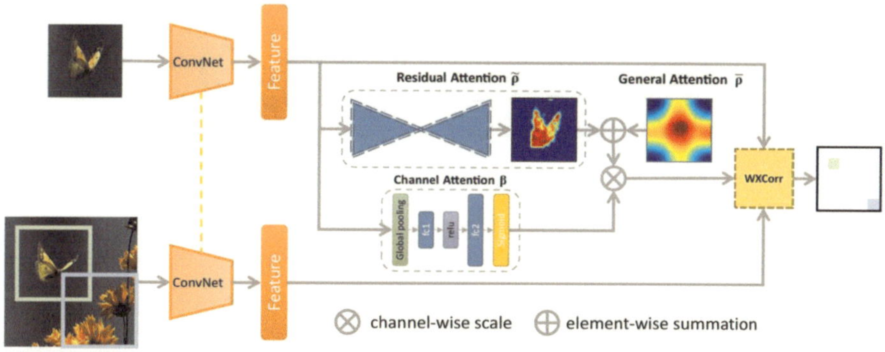

Figure 7.
RASNet tracking framework [13].

network to encode and learn more adaptive target representation from multiple levels. It also investigates a variety of attentional techniques to adjust offline feature representation learning to track a specific target. All training operations in RASNet are completed during the offline phase to ensure efficient tracking performance. Tracker's general attention mechanism gradually converges in the center, which is similar to a Gaussian distribution. It represents the center position as a more important part of the training samples than the peripheral parts, which is tremendously beneficial to train the siamese network. A residual attention module is incorporated to improve the general attention module performance and combinedly called the dual attention (DualAtt) model. The residual module helps to learn better representation and reduces bias on the training data. Furthermore, the channel attention module integrates to a single branch of the siamese network to improve the network discrimination ability, which learns through channel-wise features.

- SCSAtt [10]: Stacked channel-spatial attention learning (SCSAtt) employed channel attention and spatial attention mechanisms together. Channel attention uses to learn "what" information, and spatial attention focuses on the location information by learning "where" information of the target. To improve tracking performance with end-to-end learning, SCSAtt combines "what" and "where" information modules and focuses on the most nontrivial part of the object. **Figure 3** shows the channel attention and spatial attention mechanisms. **Figure 8** illustrates the SCSAtt tracker combining channel attention and spatial attention. The overall framework tries to balance the tracker's accuracy (success and precision) and speed. SCSAtt extends the baseline siamese network by incorporating the stacked channel-spatial attention in the target branch to handle challenges. SCSAtt channel attention and spatial attention modules consider max-pooled and global average-pooled features together to learn better target representation and discrimination learning. These improved features help the network to locate and identify the target in challenging scenarios, such as background clutter, fast motion, motion blur, and scale variation. SCSAtt does not employ any updating mechanisms in the tracking framework and considers only a pretrained model during testing, which helps to ensure fast tracking performance.

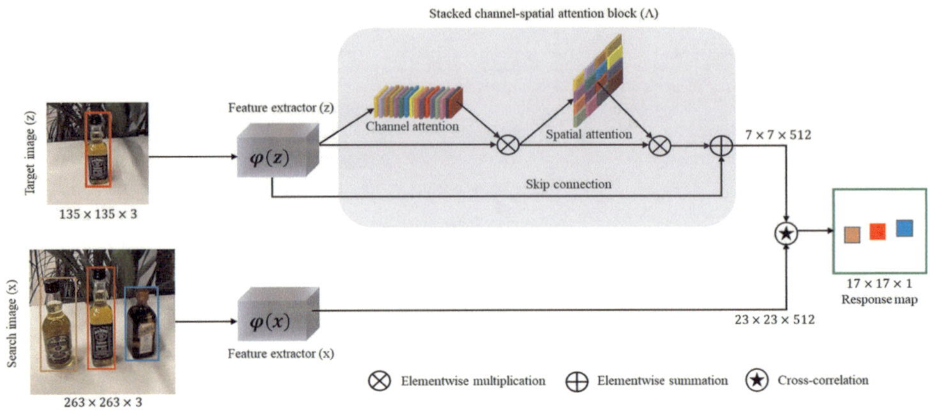

Figure 8.
SCSAtt tracking framework [10].

4. Experimental analysis and results

This section describes the experimental analysis and compares the results of the visual trackers over the OTB benchmark. The most popular comparison on the OTB benchmark is the OTB2015 benchmark [17, 18]. It is also familiarized as the OTB100 benchmark because of consisting 100 challenging video sequences for evaluating tracking performance. Besides, the subset of OTB100 benchmark named OTB50 benchmark is also considered for evaluating tracking performance. It contains the most challenging 50 sequences among hundred sequences. The OTB video sequences are categorized into 11 challenging attributes, such as scale variation (SV), background clutter (BC), fast motion (FM), motion blur (MB), low resolution (LR), in-plane rotation (IPR), out-plane rotation (OPR), deformation (DEF), occlusion (OCC), illumination variation (IV), and out-of-view (OV).

Usually, one-pass evaluation (OPE) uses to compute success and precision plots. The percentage of overlap score between the predicted and ground-truth bounding box is considered as success scores. The center location error of the predicted and ground-truth bounding box is considered as precision scores. The overlap score is computed by the intersection over union (IOU), and the center location error is computed by the center pixel distance. Success plots and precision plots are drawn using the tracking community-provided OTB toolkit based on these two scores. The precision and success plots thresholds are 20 pixels distance and 0.5 IOU score, respectively, and considered accurate tracking. The following subsections demonstrate a quantitative and qualitative analysis by comparing the tracking speed.

4.1 Quantitative and qualitative comparison and analysis

To compute a fair comparison, we carefully selected various trackers including attentional and non-attentional siamese-based trackers. **Figures 9** and **10** show the compared trackers' results on the OTB100 and OTB50 benchmarks, respectively. The compared trackers in **Figures 9** and **10** are siamese-based. Among them, SA-Siam [11], SCSAtt [10], MemDTC [35], MemTrack [34], and SiamFRN [12] utilize attention mechanism to improve the baseline SiamFC tracker [9]. SiamFC achieves 77.1% and 69.2% for overall precision plots, and 58.2% and 51.6% for overall success plots on OTB100 and OTB50 benchmarks. The attention-based tracker SA-Siam shows the

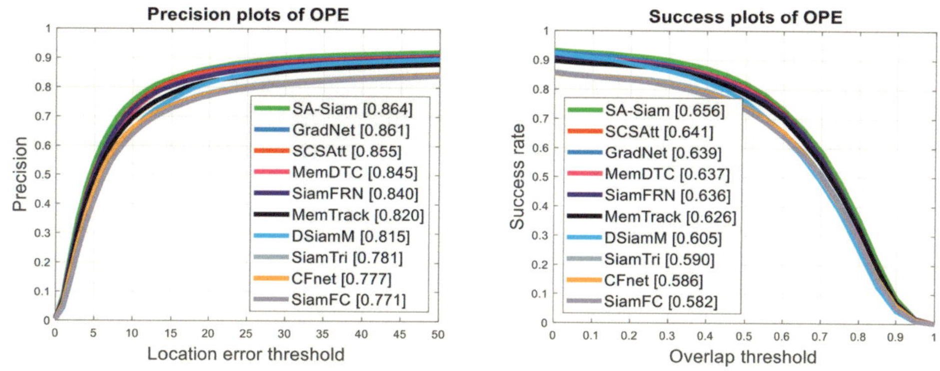

Figure 9.
Compared trackers' results on OTB100 benchmark.

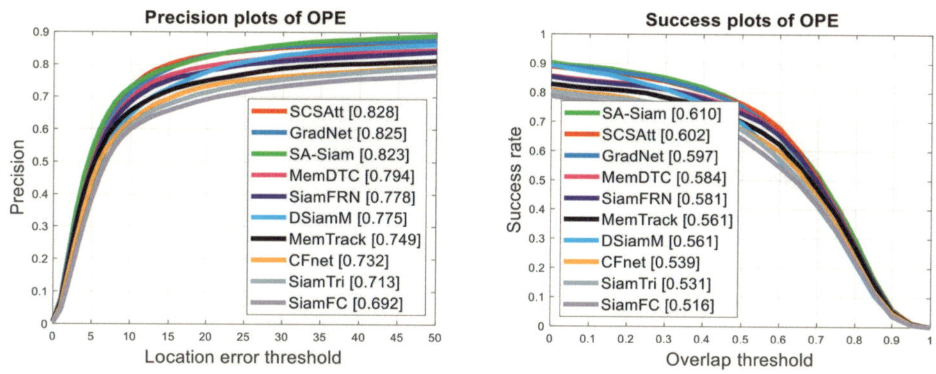

Figure 10.
Compared trackers' results on OTB50 benchmark.

dominating performance among the compared trackers. It acquires 86.4% and 65.6% precision and success scores on the OTB100 benchmark, respectively. The OTB50 benchmark also achieves 82.3% in the precision score and 61.0% in the success score.

The overall performance of the attention-integrated siamese trackers is higher than other siamese-based trackers. Among the other siamese trackers, GradNet performance is better due to its expensive tracking time operation. GradNet performs 86.1% and 82.5% for precision plots, and 63.9% and 59.7% success plots on OTB100 and OTB50 benchmarks. The other siamese-based trackers', including DSiamM, SiamTri, and CFnet, performance is not much improved than the original siamese pipeline. However, the attention with the siamese baseline tracker shows improving the tracker's overall performance. The attention-integrated siamese trackers, including SCSAtt and SiamFRN, utilize the same channel attention mechanism inside their framework. They achieve 82.8% and 77.8% for precision, and 60.2% and 58.1% for precision success plots, respectively, on the OTB50 benchmark. The trackers with the LSTM attention network (MemDTC and MemTrack) also performed better than the baseline siamese tracker. Both follow a similar attention mechanism except considering different features for memory, which makes the performance difference. MemDTC achieves 84.5% for precision plots, which is 2.5% higher than the MemTrack scores (82.0%). Similarly, the gap between them is 1.1% for success scores on the OTB100 benchmark. MemDTC also performs better than MemTrack on the OTB50 benchmark.

Figures 11 and **12** show the trackers' performance comparison on the challenging attributes of the OTB100 benchmark in terms of precision and success plots. For better visualization of these two figures, the interested reader may check this link:

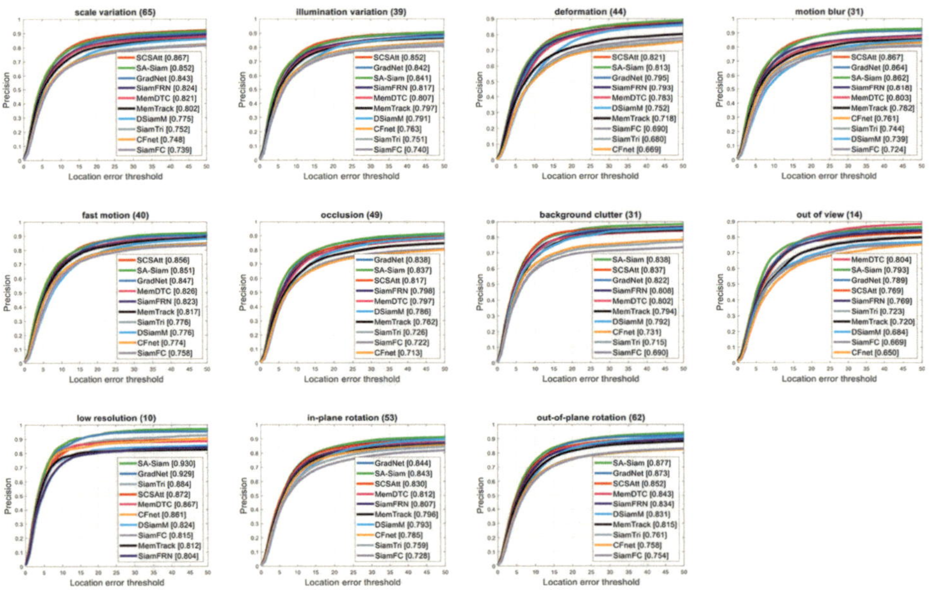

Figure 11.
Compared trackers' performance on the challenging attributes of OTB100 benchmark in terms of precision plots.

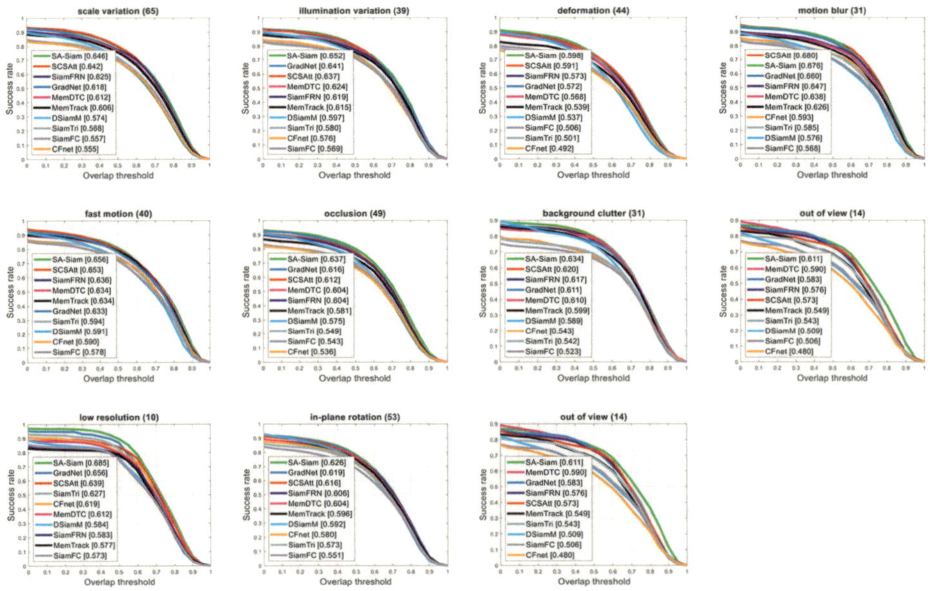

Figure 12.
Compared trackers' performance on the challenging attributes of OTB100 benchmark in terms of success plots.

https://github.com/maklachur/VOT_Book-Chapter. SCSAtt tracker performs better in precision plots than other trackers in several challenging scenarios, such as scale variation, illumination variation, deformation, motion blur, and fast motion. SCSAtt utilizes channel attention and spatial attention mechanism into the baseline SiamFC model. Furthermore, the channel attention-based SA-Siam tracker performs better than the other siamese-based trackers, including CFnet, DSiamM, and SiamTri. SA-Siam also shows the dominating performance on other trackers over the OTB100 benchmark in the success plots of challenging attributes. It performs better than the other trackers except on the motion blur challenge, whereas SCSAtt performs better than the other trackers for success plots.

Figure 13 illustrates the qualitative comparison results among trackers over several challenging sequences from the OTB100 benchmark. For better visualization of this result, the interested reader may check this link: https://github.com/maklachur/VOT_Book-Chapter. The overall tracking accuracy of attention-based trackers is better than the other trackers. They can track the target object more correctly with accurate bounding boxes from the background information. We observed that most trackers fail to handle the target in the car sequence, but MemTrack and MemDTC trackers manage to provide better tracking. Similarly, SCSAtt, SA-Siam, and SiamFRN show accurate tracking for other compared sequences, whereas the non-attentional trackers suffer handling the target accurately.

4.2 Speed comparison and analysis

In order to compare tracking speed, we selected trackers from our previous comparison for quantitative and qualitative analysis. **Table 1** shows the speed comparison results in terms of FPS and corresponding success and precision scores on the OTB100 benchmark. We observed that SiamFC tracking speed (86 FPS) shows the highest tracking speed, but it achieves the lowest accuracy scores in terms of success and precision. Therefore, it could not utilize its full potential of tracking speed. The motivation of designing trackers is not just to improve the tracking speed, but they

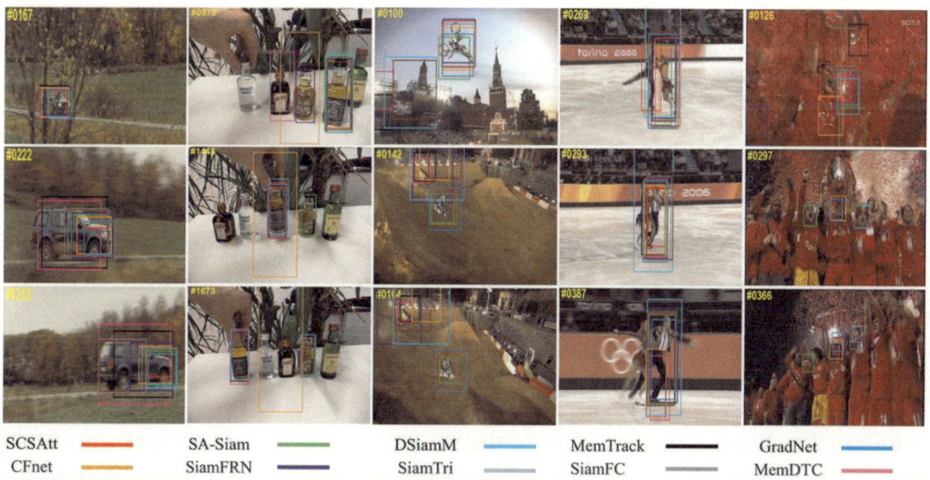

Figure 13.
The qualitative comparison results among trackers over several challenging sequences (carScale, liquor, motorRolling, skating2–2, and soccer) from the OTB100 benchmark.

Tracker name	Speed (FPS)	Success score (%)	Precision score (%)
SA-Siam [11]	50	65.6	86.4
RASNet [13]	83	64.2	—
SCSAtt [10]	61	64.1	85.5
Img-Siam [14]	50	63.8	84.6
SimaFRN [12]	60	63.6	84.0
MemDTC [35]	40	63.7	84.5
MemTrack [34]	50	62.6	82.0
SiamFC [9]	86	58.2	77.1

The red highlight represents the best, green represents the second best, and blue represents the third-best performance.
**RASNet paper did not provide the precision score that is why we do not include it in our comparison.*

Table 1.
The speed comparison results in terms of FPS and corresponding success and precision scores on the OTB100 benchmark.

should be able to track the target in challenging scenarios. Preserving a balance between speed and accuracy is essential when designing a tracker for real-time applications. Most of the presented trackers in our comparison illustrate better performance than the SiamFC. RASNet and SCSAtt achieve the second-highest and third-highest tracking speeds, respectively. They also show better accuracy on success scores and show a balance performance.

Most trackers presented in **Table 1** show the high tracking speed because of leveraging the SiamFC pipeline and computing template image only for the very first frame of the video sequence. However, MemDTC achieves the lowest tracking speed among the other trackers, which is 40 FPS. It utilizes the memory mechanism for updating the target template during tracking, which reduces its operational efficiency. SA-Siam, Img-Siam, MemTrack, and SiamFRN achieve 50 FPS, 50 FPS, 50 FPS, and 60 FPS tracking speed, respectively. The motivation of these trackers is maintaining a balance between the tracking speed and accuracy utilizing the siamese tracking framework for handling challenging sequences fully.

5. Conclusion and future directions

The attention mechanism is very simple but powerful for improving the network learning ability. It is beneficial for better target representation and enhancing tracker discrimination ability with fewer parameter overhead. The baseline siamese tracker does not perform well on accuracy on the challenging scenarios due to its insufficient feature learning and distinguishing inability between foreground and background. The attention mechanism is integrated into the baseline tracker pipeline to overcome the underlying siamese issues and improve the tracking performance. Attention helps to prioritize the features by calculating the relevant weights gain of the individual feature map. Therefore, it learns to highlight the important features of the target, which helps to handle challenges during tracking. In our study, we present a detailed discussion about the attention embedding in siamese trackers. The attention-based siamese trackers show outstanding performance and domination over other non-attentional trackers in the compared results. For example, SA-Siam and SCSAtt achieve high tracking accuracy in success and precision plots on most challenging attributes, representing the robustness of the model.

Furthermore, we observed that the employed attention mechanism in the target branch performs well instead of integrating only in the search branch or both branches. Besides this, multiple attention mechanisms are considered rather than the single attention mechanism to focus on the target class and the location information. Since the location information is important for correctly predicting the object's bounding box, the spatial information-focused module helps to improve the tracker's effectiveness on challenges. RASNet and SCSAtt trackers used the multiple attention mechanisms in their pipeline to handle the challenging sequences. The trackers' performance on challenging attributes in **Figures 11** and **12** proves the attention mechanism advantages. Using the attention mechanisms inside the tracker framework would be a better choice for future tracker developments. Therefore, improving the overall tracker performance on challenges and preserving the balance performance between accuracy and speed, integrating attention mechanisms are recommended for designing the future tracking framework.

Acknowledgements

This research was supported by Basic Science Research Program through the National Research Foundation of Korea (NRF) funded by the Ministry of Education, Science and Technology (NRF-2019R1A2C1010786).

Author details

Md. Maklachur Rahman* and Soon Ki Jung
Virtual Reality Lab, School of Computer Science and Engineering, Kyungpook National University, South Korea

*Address all correspondence to: maklachur@gmail.com; maklachur@knu.ac.kr

References

[1] Attard L, Farrugia RA. Vision based surveillance system. In: 2011 IEEE EUROCON-International Conference on Computer as a Tool. IEEE; 2011. pp. 1-4

[2] Janai J, Güney F, Behl A, Geiger A. Computer vision for autonomous vehicles: Problems, datasets and state-of-the-art. arXiv preprint arXiv: 170405519. 2017;**12**:1-308

[3] Lu WL, Ting JA, Little JJ, Murphy KP. Learning to track and identify players from broadcast sports videos. IEEE Transactions on Pattern Analysis and Machine Intelligence. 2013;**35**(7): 1704-1716

[4] Pavlovic VI, Sharma R, Huang TS. Visual interpretation of hand gestures for human-computer interaction: A review. IEEE Transactions on Pattern Analysis and Machine Intelligence. 1997; **19**(7):677-695

[5] Song Y, Ma C, Gong L, Zhang J, Lau RW, Yang MH. Crest: Convolutional residual learning for visual tracking. In: Proceedings of the IEEE International Conference on Computer Vision. Italy: IEEE; 2017. pp. 2555-2564

[6] Danelljan M, Robinson A, Khan FS, Felsberg M. Beyond correlation filters: Learning continuous convolution operators for visual tracking. In: European Conference on Computer Vision. Netherland: Springer; 2016. pp. 472-488

[7] Danelljan M, Bhat G, Shahbaz Khan F, Felsberg M. Eco: Efficient convolution operators for tracking. In: Proceedings of the IEEE Conference on Computer Vision and Pattern Recognition. Hawaii: IEEE; 2017. pp. 6638-6646

[8] Nam H, Han B. Learning multi-domain convolutional neural networks for visual tracking. In: Proceedings of the IEEE Conference on Computer Vision and Pattern Recognition. Nevada: IEEE; 2016. pp. 4293-4302

[9] Bertinetto L, Valmadre J, Henriques JF, Vedaldi A, Torr PH. Fully-convolutional siamese networks for object tracking. In: European Conference on Computer Vision. Netherland: Springer; 2016. pp. 850-865

[10] Rahman MM, Fiaz M, Jung SK. Efficient visual tracking with stacked channel-spatial attention learning. IEEE Access. Utah: IEEE. 2020;**8**:100857-100869

[11] He A, Luo C, Tian X, Zeng W. A twofold siamese network for real-time object tracking. In: Proceedings of the IEEE Conference on Computer Vision and Pattern Recognition. Utah: IEEE; 2018. pp. 4834-4843

[12] Rahman M, Ahmed MR, Laishram L, Kim SH, Jung SK, et al. Siamese high-level feature refine network for visual object tracking. Electronics. 2020;**9**(11): 1918

[13] Wang Q, Teng Z, Xing J, Gao J, Hu W, Maybank S. Learning attentions: Residual attentional siamese network for high performance online visual tracking. In: Proceedings of the IEEE Conference on Computer Vision and Pattern Recognition. Utah: IEEE; 2018. pp. 4854-4863

[14] Qin X, Fan Z. Initial matting-guided visual tracking with Siamese network. IEEE Access. 2019;**03**:1

[15] Fiaz M, Rahman MM, Mahmood A, Farooq SS, Baek KY, Jung SK. Adaptive feature selection Siamese networks for visual tracking. In: International

Workshop on Frontiers of Computer Vision. Japan: Springer; 2020. pp. 167-179

[16] Woo S, Park J, Lee JY, So KI. Cbam: Convolutional block attention module. In: Proceedings of the European Conference on Computer Vision (ECCV). Germany: Springer; 2018. pp. 3-19

[17] Wu Y, Lim J, Yang MH. Object tracking benchmark. IEEE Transactions on Pattern Analysis and Machine Intelligence. 2015;**37**(9): 1834-1848

[18] Wu Y, Lim J, Yang MH. Online object tracking: A benchmark. In: Proceedings of the IEEE Conference on Computer Vision and Pattern Recognition. Oregon: IEEE; 2013. pp. 2411-2418

[19] Bromley J, Guyon I, LeCun Y, Säckinger E, Shah R. Signature verification using a "siamese" time delay neural network. In: Advances in Neural Information Processing Systems. US: NIPS; 1994. pp. 737-744

[20] Tao R, Gavves E, Smeulders AW. Siamese instance search for tracking. In: Proceedings of the IEEE Conference on Computer Vision and Pattern Recognition. Nevada: IEEE; 2016. pp. 1420-1429

[21] Held D, Thrun S, Savarese S. Learning to track at 100 fps with deep regression networks. In: European Conference on Computer Vision. Netherland: Springer; 2016. p. 749–765

[22] Chen K, Tao W. Once for all: A two-flow convolutional neural network for visual tracking. IEEE Transactions on Circuits and Systems for Video Technology. 2018;**28**(12): 3377-3386

[23] Valmadre J, Bertinetto L, Henriques J, Vedaldi A, Torr PH. End-to-end representation learning for correlation filter based tracking. In: Proceedings of the IEEE Conference on Computer Vision and Pattern Recognition. Hawaii: IEEE; 2017. pp. 2805-2813

[24] Dong X, Shen J. Triplet loss in siamese network for object tracking. In: Proceedings of the European Conference on Computer Vision (ECCV). Germany: Springer; 2018. pp. 459-474

[25] Guo Q, Feng W, Zhou C, Huang R, Wan L, Wang S. Learning dynamic siamese network for visual object tracking. In: Proceedings of the IEEE International Conference on Computer Vision. Italy: IEEE; 2017. pp. 1763-1771

[26] Morimitsu H. Multiple context features in Siamese networks for visual object tracking. In: Proceedings of the European Conference on Computer Vision (ECCV). Germany: Springer; 2018

[27] Khan FS, Van de Weijer J, Vanrell M. Modulating shape features by color attention for object recognition. International Journal of Computer Vision. IJCV: Springer; 2012;**98**(1):49-64

[28] Fu J, Liu J, Tian H, Li Y, Bao Y, Fang Z, et al. Dual attention network for scene segmentation. In: Proceedings of the IEEE Conference on Computer Vision and Pattern Recognition. 2019. pp. 3146-3154

[29] Xu J, Zhao R, Zhu F, Wang H, Ouyang W. Attention-aware compositional network for person re-identification. In: Proceedings of the IEEE Conference on Computer Vision and Pattern Recognition. Utah: IEEE; 2018. pp. 2119-2128

[30] Li D, Wen G, Kuai Y, Porikli F. End-to-end feature integration for correlation

filter tracking with channel attention. IEEE Signal Processing Letters. SPL: IEEE; 2018;**25**(12):1815-1819

[31] Fiaz M, Mahmood A, Baek KY, Farooq SS, Jung SK. Improving object tracking by added noise and channel attention. Sensors. Utah: IEEE; 2020; **20**(13):3780

[32] Rahman MM. A DWT, DCT and SVD based watermarking technique to protect the image piracy. arXiv preprint arXiv:13073294. 2013

[33] Rahman MM, Ahammed MS, Ahmed MR, Izhar MN. A semi blind watermarking technique for copyright protection of image based on DCT and SVD domain. Global Journal of Research In Engineering. SPL: IEEE; 2017;**16**

[34] Yang T, Chan AB. Learning dynamic memory networks for object tracking. In: Proceedings of the European Conference on Computer Vision (ECCV). Germany: Springer; 2018. pp. 152-167

[35] Yang T, Chan AB. Visual tracking via dynamic memory networks. IEEE Transactions on Pattern Analysis and Machine Intelligence. TPAMI: IEEE; 2019

[36] Zheng Z, Wu W, Zou W, Yan J. End-to-End Flow Correlation Tracking with Spatial-Temporal Attention. Utah: IEEE; 2018. pp. 548-557

[37] Krizhevsky A, Sutskever I, Hinton GE. Imagenet classification with deep convolutional neural networks. In: Advances in Neural Information Processing Systems. US: NIPS; 2012. pp. 1097-1105

Chapter 3

Robust Template Update Strategy for Efficient Visual Object Tracking

Awet Haileslassie Gebrehiwot, Jesus Bescos and Alvaro Garcia-Martin

Abstract

Real-time visual object tracking is an open problem in computer vision, with multiple applications in the industry, such as autonomous vehicles, human-machine interaction, intelligent cinematography, automated surveillance, and autonomous social navigation. The challenge of tracking a target of interest is critical to all of these applications. Recently, tracking algorithms that use siamese neural networks trained offline on large-scale datasets of image pairs have achieved the best performance exceeding real-time speed on multiple benchmarks. Results show that siamese approaches can be applied to enhance the tracking capabilities by learning deeper features of the object's appearance. SiamMask utilized the power of siamese networks and supervised learning approaches to solve the problem of arbitrary object tracking in real-time speed. However, its practical applications are limited due to failures encountered during testing. In order to improve the robustness of the tracker and make it applicable for the intended real-world application, two improvements have been incorporated, each addressing a different aspect of the tracking task. The first one is a data augmentation strategy to consider both motion-blur and low-resolution during training. It aims to increase the robustness of the tracker against a motion-blurred and low-resolution frames during inference. The second improvement is a target template update strategy that utilizes both the initial ground truth template and a supplementary updatable template, which considers the score of the predicted target for an efficient template update strategy by avoiding template updates during severe occlusion. All of the improvements were extensively evaluated and have achieved state-of-the-art performance in the VOT2018 and VOT2019 benchmarks. Our method (VPU-SiamM) has been submitted to the VOT-ST 2020 challenge, and it is ranked 16th out of 38 submitted tracking methods according to the Expected average overlap (EAO) metrics. VPU_SiamM Implementation can be found from the VOT2020 Trackers repository[1].

Keywords: real-time, tracking, template update, Siamese

1. Introduction

Visual object tracking (VOT), commonly referred to as target tracking, is an open problem in computer vision; this is due to a broad range of possible applications and

[1] https: //www.votchallenge.net/vot2020/trackers. html

potential tracking challenges. Thus, it has been divided into sub-challenges according to several factors, which include: the number of targets of interest, the number of cameras, the type of data (i.e., medical, depth, thermal, or RGB images), static or moving camera, offline or online (real-time) processing.

Visual object tracking is the process of estimating and locating a target over time in a video sequence and assigning a consistent label to the tracked object across each video sequence frame. VOT algorithms have been utilized as a building block in more complex applications of computer vision such as traffic flow monitoring [1], human-machine interaction [2], medical systems [3], intelligent cinematography [4], automated surveillance [5], autonomous social navigation [6] and activity recognition [7]. Real-time visual target tracking is the process of locating and associating the target of interest in consecutive video frames while the action is taking place in real-time. Real-time visual target tracking plays an inevitable role in time-sensitive applications such as autonomous mobile robot control to keep track of the target of interest while the viewpoint is changing due to the movement of the target or the robot. In such a scenario, the tracking algorithm must be accurate and fast enough to detect sudden changes in the observed environment and act accordingly to prevent losing track of the quickly moving target of interest.

Since the start of the Visual-Object-Tracking(VOT) Real-time challenge in 2017, Siamese network-based tracking algorithms have achieved top performance and won in the VOT real-time challenge with a considerable margin over the rest of the trackers. Nearly all top ten trackers applied the siamese network, and also the winners. The dominant methodology in real-time tracking, therefore, appears to be associated. A siamese network aims to learn a similarity function. It has a Y-shaped network architecture that takes two input images and returns similarity as an output. Siamese networks are utilized to compare the similarity between the template and the candidate images to determine if the two input images have an identical pattern(similarity). In the past few years, a series of state-of-the-art siamese-based trackers have been proposed, and all of them utilize embedded features by employing CNN to compute similarity and produce various types of output, such as similarity score(probability measure), response map(two-dimensional similarity score map), and bounding box location of the target.

Luca Bertinetto et al. [8] proposed Siamese fully convolutional network (SiameseFC) to addresses the broad similarity learning between a target image and

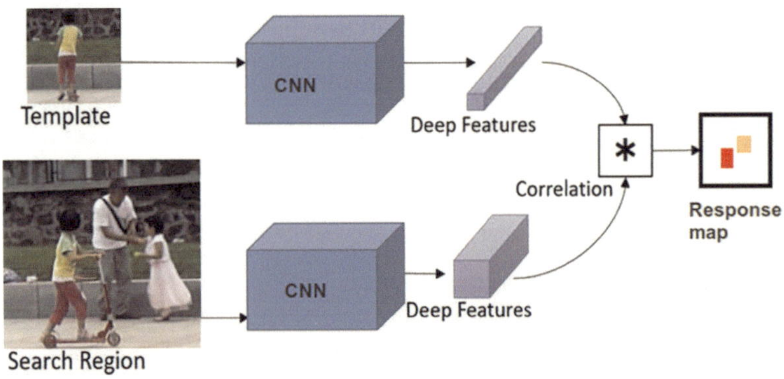

Figure 1.
Fully-convolutional Siamese architecture. The output is a scalar-valued score map whose dimension depends on the size of the search image [8].

search image, as presented in **Figure 1**. According to the VOT winner rules, the winning real-time tracker of the VOT2017 [9] was SiamFC. SiamFC applies a fully-convolutional siamese network trained offline to locate an exemplar (template) image inside a larger search image **Figure 1**. The network is fully convolutional w.r.t search image: dense and efficient sliding window evaluation is achieved with a bilinear layer that computes the cross-correlation of two inputs. The deep convolutional network is trained offline with" ILSVRC VID" dataset [10] to address a general similarity learning problem and maximize target discrimination power. During tracking, SiamFC takes two images and infers a response map using the learned similarity function. The new target position is determined at the maximum value on the response map, where it depicts a maximum similarity **Figure 1**. As improvement in Siamese based tracking methods, Qiang Wang et al. [11] proposed SiamMask aiming to improve the ability of the SiamFC network to differentiate between the background and the foreground by augmenting their loss with a binary segmentation task. SiamMask is a depth-wise cross-correlation operation performed on a channel-by-channel basis, to keep the number of channels unchanged. The result of the depth-with cross-correlation indicated as RoW (response of candidate window), then distributed into three branches, respectively segmentation, regression, and classification branches **Figure 2**.

Seven of the top ten realtime trackers (SiamMargin [12], SiamDWST [13], SiamMask [11], SiamRPNpp [14], SPM [15] and SiamCRF-RT) are based on siamese correlation combined with bounding box regression. In contrast, the top performers of the VOT2019 Real-time challenge are from the class of classical siamese correlation trackers, and siamese trackers with region proposals [16]. Although these methods showed a significant improvement, there was small attention on how to carefully update the template of the target as time goes from the start of the tracking. In all top performers, the target template is initialized in the first frame and then kept fixed during the tracking process. However, diverse variations regarding the target usually occur in the process of tracking, i.e., camera orientation, illumination change, self-rotation, self-deformation, scale, and appearance change. Thus, failing to update the target template leads to the early failure of the tracker. In such scenarios, it is crucial to adapt the target template model to the current target appearance. In addition to this, most of the tracking methods fail when motion-blurred frames or frames with low-resolution appear in the video sequence, as depicted in **Figures 3** and **4**. We

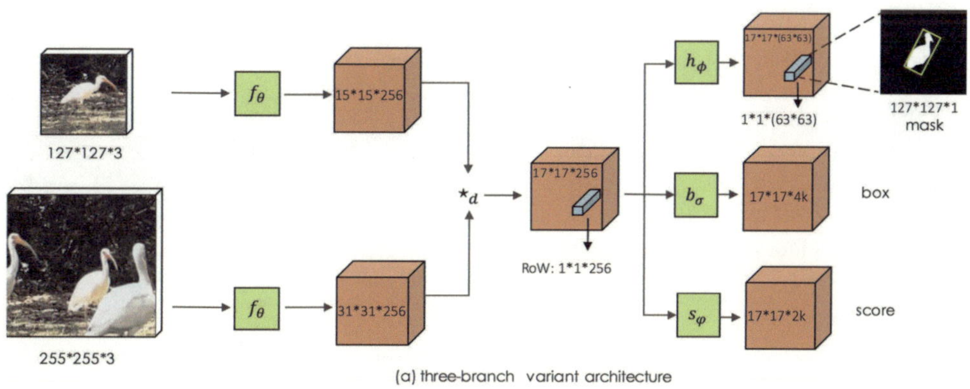

Figure 2.
An illustration of SiamMask with three branches, respectively segmentation, regression, and classification branches; where $*_d$ denotes depth-wise cross correlation [11].

believe that this case arguably arises from the complete lack of similar training samples. Therefore one must incorporate a data-augmentation strategy to consider both motion-blur and low-resolution during training to significantly increase the diversity of datasets available for training without actually gathering new data.

2. Method

The problem of establishing a correspondence between a single target in consecutive frames can be affected by factors such as initializing a track, updating it robustly, and ending the track. The tracking algorithm receives an input frame from the camera module and performs the visual tracking over a frame following a siamese network-based tracking approach. Since developing a new tracking algorithm from scratch is beyond the scope of this chapter, a state-of-the-art siamese-based tracking algorithm called siammask [11], one of the top performers in the VOT2019 real-time challenge, is used as a backbone of our tracking algorithm.

To mitigate the limitations associated with Siamese-based tracking methods. This section presents two improvements on top of the SiamMask implementation. *VPU_SiamM Implementation can be found from VOT2020 Trackers repository*[2].

2.1 Data-augmentation

As mentioned in the introduction, the siamese-based tracker fails when motion-blurred frames or frames with low-resolution appear in the video sequence, as depicted in **Figures 3** and **4**. Therefore to address the problems, a tracking algorithm should incorporate a data-augmentation strategy to consider both motion-blur and low-resolution during training. Since data augmentation is a strategy that significantly increases the diversity of datasets available for training without actually gathering new data, it will require implementing the data augmentation techniques explained through the following sub-sections.

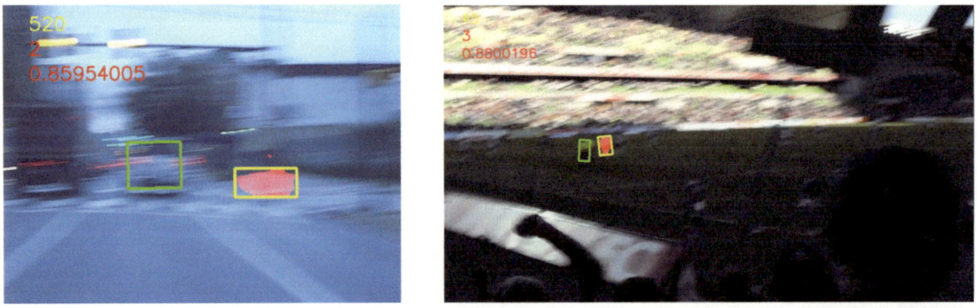

(a) Tracking failure from car VOT2019 dataset

(b) Tracking failure from the soccer VOT-2019 dataset

Figure 3.
An example of SiamMask failure due to motion-blur, green and yellow bounding box indicates ground truth and predicted target respectively.

[2] https: //www.votchallenge.net/vot2020/trackers. html

DOI: http://dx.doi.org/10.5772/intechopen.101800

(a) Tracking failure from agility VOT2019 dataset

(b) Tracking failure from handball VOT-2019 dataset

Figure 4.
An example of SiamMask failure due to low resolution, green and yellow bounding box indicates ground truth and predicted target respectively.

2.1.1 Data-augmentation for motion-blur

Kernel filters are a prevalent technique in image processing to blur images. These filters work by sliding an $n \times n$ matrix across an image with a Gaussian blur filter, resulting in a blurry image. Intuitively, blurring images for data augmentation could lead to higher resistance to motion blur during testing [17]. **Figure 5** illustrates an example of a motion-blurred frame generated by the developed data-augmentation technique.

2.1.2 Data-augmentation for low-resolution

We followed a Zhangyang Wang et al. [18] approach to generate a low-resolution dataset. During training, the original (High Resolution) images are first downscaled by $scale = 4$ and then upscaled back by $scale = 4$ with nearest-neighbor interpolation as low-resolution images. A small additive Gaussian noise is added as a default data augmentation during training. An illustrates of a low-resolution frame generated by the developed data-augmentation technique is depicted in **Figure 6**.

Figure 5.
An example of motion blurred frame (left image) generated from original frame (right image) using the developed data-augmentation for motion-blur technique.

(a) Original image. (b) down-scaled by scale=4. (c) up-scaled back by scale=4.

Figure 6.
An illustrates on how the low-resolution data augmentation generation are performed (from (a) to (c)).

2.2 Target template updating strategy

The target template update mechanism is an essential step, and its robustness has become a crucial factor influencing the quality of the tracking algorithm. To tackle this problem, more recent Siamese trackers [19–21] have implemented a simple linear update strategy using a running average with a constant learning rate. However, A simple linear update is often inadequate to cope with the changes needed and to generalize to all potentially encountered circumstances. Lichao Zhang et al. [22] proposes to replace the hand-crafted update function with a method that learns to update, using a convolutional neural network called *UpdateNet*, aims to estimate the optimal template for the next frame. However, excessive reliance on a single updated template may suffer from catastrophic drift and the inability to recover from tracking failures.

One can argue the importance of the original initial and supplementary updatable templates, which incorporate the up-to-date target information. To this end, we have incorporated a template updates strategy that utilizes both the initial template (ground truth template)T_G and an updatable template T_i. Consequently, the initial template T_G provides highly reliable information. It increases robustness against model drift, whereas an updatable template T_i integrates the new target information at the predicted target location given by the current frame. However, when a target is temporarily occluded, such as when a motorbike passes through the forest and is shielded by trees **Figure** 7, updating the template during occlusion is not required as it may cause template pollution because of shield interference. Therefore, our system needs to recognize if occlusion occurs and be able to decide whether to update the template or not. Examples of occlusion in tracking are shown in **Figures** 7 and **8**. As depicted in **Figures** 7 and **8**, when the target is occluded, the score becomes small, indicating the similarity between the tracked target in the current frame and the

(a) No occlusion, score: 0.995 (b) Partial occlusion, score: 0.862 (c) Full occlusion, score: 0.446 (d) No occlusion, score: 0.992

Figure 7.
Overview on how the target similarity score (red) varies under different occlusion scenario during tracking process. The similarity score is indicated in red color in the top left of each frame, VOT2019 road dataset. Where blue: Ground truth, red: Tracking result.

(a) No occlusion, score: 0.993 (b) Full occlusion, score: 0.621 (c) Partial occlusion, score: 0.883 (d) No occlusion, score: 0.995

Figure 8.
Overview on how the target similarity score varies under different occlusion scenario during tracking process, VOT2019 girl dataset. Where blue: Ground truth, red: Tracking result.

template is low. Thus the response on the score value can be used as the criterion for strategic decision. **Figure 9** illustrates an overview of the method.

2.2.1 Updating with a previous and a future template

In 2.2 the target template update strategy considers the target appearance only from the previous frame. However, in this section, we introduce an alternative template update strategy that considers both the target appearance from the previous frame and the target appearance in the future frame, which incorporates future information of the target appearance by updating the updatable template T_i described in 2.2. The template updating mechanism is shown in **Figure 10**. During online tracking, the template updating and the tracking procedure works as follows:

1. Tracking procedure on the next frame $i + 1$ is applied using both the previously updated target template T_i and the ground truth target template T_G.

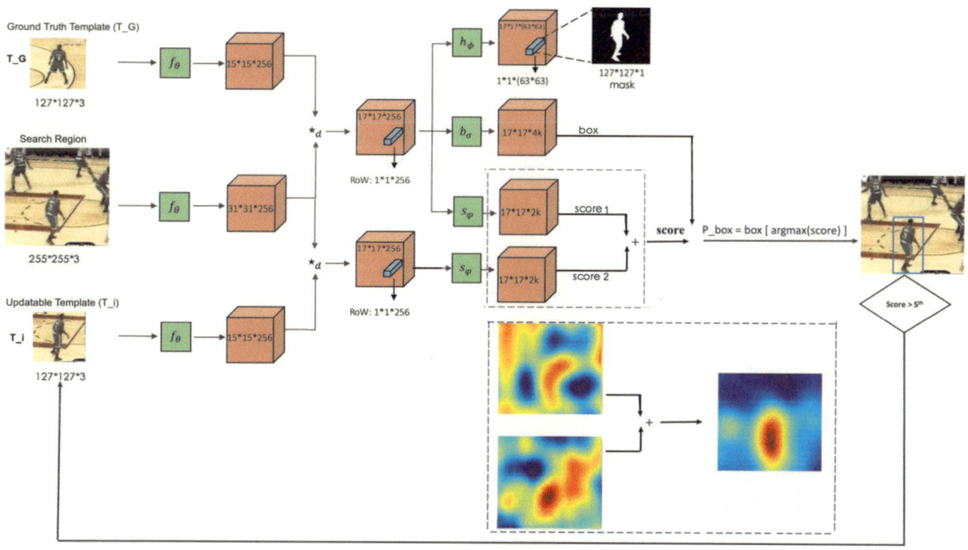

Figure 9.
Target template update strategy: Where T_G is the ground truth template, T_i is an updatable template, S^{th} is the score threshold and P_box is the predicted target location.

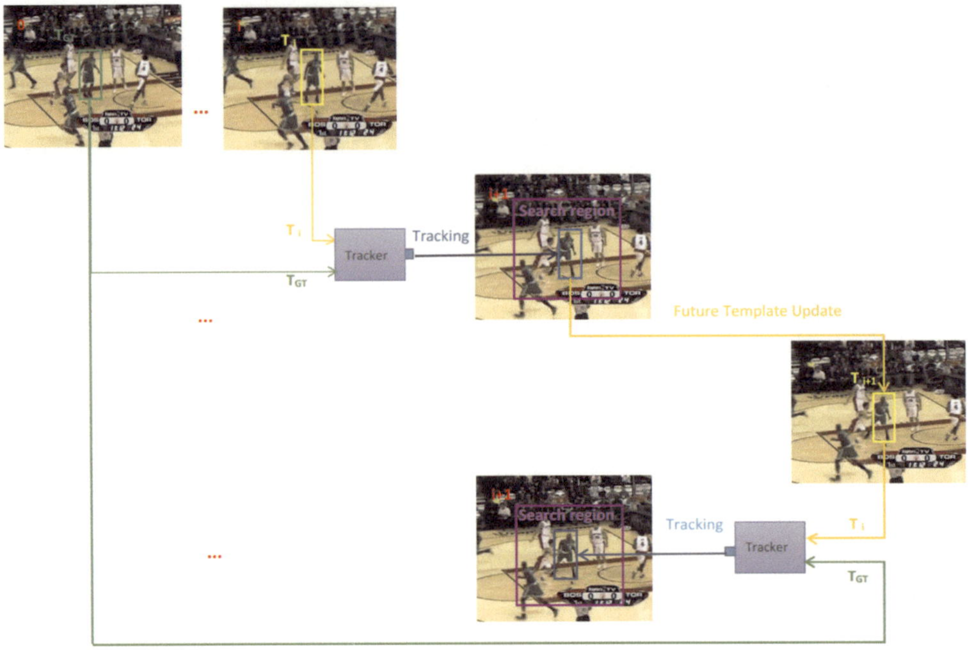

Figure 10.
Updating with a previous and a future template, where T_G is ground truth target template, T_i is previous target template and T_i + 1 is future target template. Where green: Ground truth template, yellow: Updatable template, blue: Tracking result.

2. Updatable template T_i is updated using the predicted target from the next frame to incorporate a piece of future information about the target.

3. Tracking procedure again on the current frame is applied using both the updated future target template T_i + 1 and the ground truth target template T_G.

First, a tracking procedure is applied using both the previous target template in T_i and the ground truth template T_G to perform tracking on the next frame. Then the updatable template T_i is updated using the predicted target on the next frame incorporating a piece of future information about the target. Finally, a tracking procedure is again applied to the current frame using both the updated future target template T_i + 1 and the ground truth template T_G.

3. Implementation

3.1 Training

The SiamMask implementation was trained using 4 Tesla V100 GPUs. In this experiment, only the refinement module of the mask branch is trained. The training process was carried out using COCO[3] and Youtube-vos[4] Datasets: The training was

[3] http://cocodataset.org/

[4] https: //youtube-vos.org/dataset/vis/

performed over ten epochs using mini-batches of 32 samples. The data augmentation techniques described in 2.1.1 and 2.1.2 were utilized for generating datasets with motion-blur and low-resolution, respectively.

3.2 Tracking

During tracking, the tracking algorithm is evaluated once per frame. The output mask is selected from the location attaining the maximum score in the classification branch and creating an optimized bounding box. Finally, the highest scoring output of the box branch is used as a reference to crop the next search frame.

3.3 Visual-object-tracking benchmark

As object tracking has gotten significant attention in the last few decades, the number of publications on tracking-related problems has made it difficult to follow the developments in the field. One of the main reasons is that there was a lack of commonly accepted annotated datasets and standardized evaluation protocols that allowed an objective comparison of different tracking methods. To address this issue, the Visual Object Tracking (VOT) workshop was organized in association with ICCV2013[5]. Researchers from the industry and academia were invited to participate in the first VOT2013 challenge, which was aimed at model-free single-object visual trackers. In contrast to related attempts in tracker benchmarking, the dataset is labeled per-frame by visual properties such as occlusion, motion change, illumination change, scale, and camera motion, offering a more systematic comparison of the trackers [23]. VOT focused on short-term tracking (no re-detection) until the VOT2017 challenge, where a new"real-time challenge" was introduced. In the Real-time challenge, the tracker constantly receives images at real-time speed. If the tracker does not respond after the new frame becomes available, the last bounding box from the previous frame is reported as the tracking result in the current frame.

3.4 VOT evaluation metrics

The VOT challenges applies a reset-based methodology. Whenever a zero overlap between the predicted bounding box and the ground truth occurs, a failure is detected, and the tracker is re-initialized five frames after the failure. There are three primary metrics used to analyze the tracking performance in visual object tracking challenge benchmark: Accuracy (A), Robustness (R), and Expected Average Overlap (EAO) [9].

3.4.1 Accuracy

Accuracy is calculated as the average overlap between the predicted and ground truth bounding boxes during successful tracking periods [23]. The tracking accuracy at time-step t is defined as the overlap between the tracker predicted bounding box A_t^T and the ground truth bounding box A_t^G

---

[5] http://www.iccv 2013.org/

$$\Phi_t = \frac{A_t^G \cap A_t^T}{A_t^G \cup A_t^T} \tag{1}$$

3.4.2 Robustness

Robustness measures how often the tracker loses/fails the target, i.e., a zero overlap between the predicted and the ground truth bounding boxes during tracking. The protocol specifies an overlap threshold to determine tracking failure. The number of failed tracked frames are then divided by the total number of frames, as depicted in Eq. (2):

$$P\tau = \frac{\{\Phi_t \le \tau\}_{k=1}^{N}}{N} \tag{2}$$

Where τ is the overlap threshold which is zero in this case, and N is the run time of the tracker in frames. A failure is identified in a frame when the overlap (as computed using Eq. (1)) is below the defined threshold τ. Thus, the robustness of the tracker is given as a normalized number of incorrectly tracked frames.

3.4.3 Expected average overlap (EAO)

For the purpose of ranking tracking algorithms, it is better to have a single metric. Thus, in 2015 the VOT challenge introduced Expected Average Overlap (EAO), which combines both Accuracy and Robustness. EAO estimates the average overlap that a tracker is expected to achieve on a large collection of short-term sequences with the same visual properties as the given dataset.

The EAO metric can be found by calculating the average of $\hat{\Phi}_{Ns}$ over typical sequence lengths, from N_{lo} to N_{hi}:

$$\Phi = \frac{1}{N_{hi} - N_{lo}} \sum_{N_s=N_{lo}}^{N_{hi}} \hat{\Phi}_{Ns} \tag{3}$$

3.5 Experiment A: Low-resolution data-augmentation

This first experiment is dedicated to evaluating the impact of the low-resolution data-augmentation technique. The data augmentation technique described in 2.1.2 was applied to generate datasets with low-resolution during the training process of the refinement module of the network.

3.5.1 Evaluation

The performance of the developed method: incorporating low-resolution datasets using data augmentation technique during training has been evaluated using the VOT evaluation metrics on the VOT2018, VOT2019 datasets. The overall Evaluation results are shown in **Table 1**.

The term # **Tracking Failures (Lost)** indicates how often the tracking algorithm lost the target in the given video sequence, basically:

- **Tracking Lost/Failure:** is when IOU between Ground-truth and Predicted Bounding box is Zero. Thus the lower the values ↓, the higher the performance.

In **Table 1**, we compare our approach against the state-of-the-art SiamMask tracker on the VOT2018 and VOT2019 benchmarks, respectively. It can be clearly observed that the data augmentation technique for incorporating low-resolution datasets has contributed to robustness improvements. The tracker's failure has decreased from 60 to 53 and from 104 to 93 in VOT2018 and VOT2019, respectively. Improvements are clearly shown especially in a video sequence with low-resolution, i.e. *handball*1, and *handball*2 as depicted in **Figure 11**.

The results obtained in **Table 1** confirm that the developed methodology significantly improved the overall performance of the tracker. This approach outperforms the original SiamMask achieving a relative gain of 2.6% and 0.4% in EAO on VOT2018 and VOT2019, respectively. Most significantly, a gain of around 3% and 5% in Robustness value has been achieved on VOT2018 and VOT2019, respectively.

3.5.2 Results

As it is depicted in **Figure 11**, The data-augmentation for incorporating low-resolution datasets during training has contributed to enhancing the tracker robustness. Thus, the tracker becomes robust against low-resolution frames during inference in relative to the original SiamMask tracker.

	VOT2018		VOT2019	
VOT Metrics	**SiamMask**	**Ours**	**SiamMask**	**Ours**
EAO ↑	0.380	**0.406**	0.280	**0.284**
Accuracy ↑	**0.609**	0.589	**0.610**	0.586
Robustness ↓	0.279	**0.248**	0.522	**0.467**

Table 1.
Comparison between SiamMask and the developed method (incorporating low-resolution data during training), under the VOT metric (EAO, Accuracy, Robustness) on VOT2018 (left) and VOT2019 (right), best results are marked in Bold.

Figure 11.
Qualitative comparison between SiamMask and developed data-augmentation technique for incorporating low-resolution datasets during training. Where blue: Ground truth, red: Tracking result.

3.6 Experiment B: Motion-blur data-augmentation

In this experiment, the data-augmentation technique for incorporating motion-blurred datasets described in 2.1.1 was applied for generating datasets with motion-blur during the training process of the refinement module of the network.

3.6.1 Evaluation

The performance of the tracking algorithm incorporating the motion-blur data augmentation technique has been evaluated using the VOT evaluation metrics on the VOT2018, VOT2019 datasets. The Overall Evaluation results are shown in **Table 2**.

The data augmentation technique for incorporating motion-blurred datasets has contributed to the overall enhancement of the tracker performance. They are clear improvements in terms of Robustness in multiple video sequences relative to SiamMask. From **Table 2**, it can be concluded that the data augmentation technique for incorporating motion-blurred datasets has contributed to the improvement in Robustness of the tracker, especially in a video sequence with a motion-blur, i.e., *ball*3, and *car*1. The overall performance of the tracker has been improved, and the developed method obtained a significant relative gain of 2.1% EAO in VOT2018 and 4% R in VOT2019, compared to the SiamMask result as it is depicted in **Table 2**.

3.6.2 Results

Figure 12 presents a visual comparison between SiamMask and the developed improvement incorporating motion-blurred datasets during training using

	VOT2018		VOT2019	
VOT Metrics	**SiamMask**	**Ours**	**SiamMask**	**Ours**
EAO ↑	0.380	**0.401**	0.280	**0.288**
Accuracy ↑	0.609	**0.610**	0.610	0.610
Robustness ↓	0.279	**0.248**	0.522	**0.482**

Table 2.
Comparison between SiamMask and the developed method (incorporating motion-blurred dataset during training), under the VOT metric (EAO, accuracy, robustness) on VOT2018 (left) and VOT2019 (right).

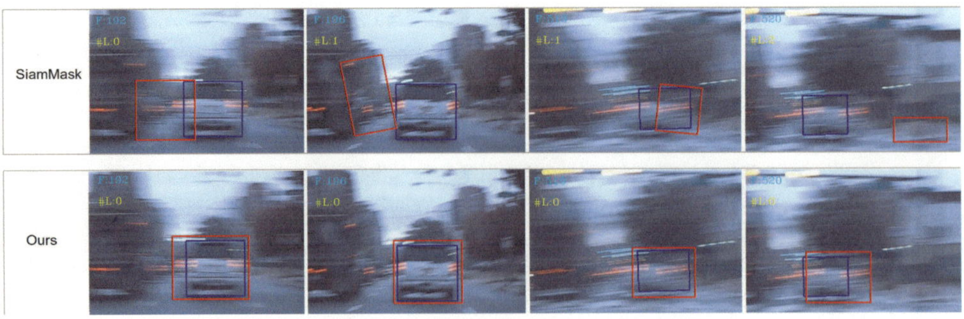

Figure 12.
Qualitative comparison between SiamMask and developed data-augmentation technique: Incorporating motion-blurred datasets during training. Where blue: Ground truth, red: Tracking result.

data-augmentation. From **Figure 12** it can be clearly observed that the data-augmentation for incorporating motion-blurred dataset during training has contributed to enhancing the tracker Robustness. Thus, the tracker has become robust against motion-blurred video frames during inference in relative to the original SiamMask tracker.

3.7 Experiment C: Target template updating strategy

When it comes to updating the target template, the question is how and when to update the target. The parameter S^{th} controls when to update the target template according to the developed template update strategy in 2.2. Thus, the 2nd (updatable) target template is updated when the predicted target's score is higher than the threshold S^{th}. Therefore determining the optimal threshold value S^{th} is the main focus in this sub experiment.

This set of experiments compares the effect of the target template updating strategy by varying the score threshold of Sth by evaluating the tracking performance on VOT2018 and VOT2019 datasets. From the experimental results shown in **Tables 3** and **4**, it can be observed that the performance of the tracker increases as the parameter S^{th} increases. Thus by using a Sth value as high as possible, we guaranty an efficient template update strategy by avoiding template updates during severe occlusion. **Figure 13** illustrates an overview of how each VOT metric (EAO, Accuracy, and Robustness) and FPS behave as we vary the Sth. Therefore, the parameter S^{th} plays an important role in deciding whether to update the target template or not when cases such as occlusion or deformation occur, as illustrated in **Figures 14** and **15**. It is worth mentioning that the template update has a negative impact on the tracker's speed since it needs to compute the feature map of the updated template for every updated template. Therefore by setting S^{th} high, we can leverage both performance and speed as it is depicted in **Figure 13**.

Figures 14 and **15** are an illustration of how the template update strategy decides when to update the updatable template. For instance in **Figure 15a** the target is not occluded; as a result the score is high, thus *Update* : *True* flag is generated indicating to update the target template, on the other hand in **Figure 15b** and **c**, the target is

		VOT-Metrics			
S^{th}	EAO ↑	Accuracy ↑	Robustness ↓	# Lost ↓	FPS ↑
0.65	0.377	0.602	0.267	57	25
0.7	0.371	0.602	0.267	57	27
0.75	0.385	0.600	0.248	53	28
0.8	0.387	0.603	0.258	55	31
0.85	0.388	**0.603**	0.243	52	32
0.9	0.393	0.602	0.239	51	35
0.95	**0.397**	0.602	**0.239**	**51**	**40**

Table 3.
Determining the optimal score threshold (S^{th}) for updating the target template under VOT-metrics on VOT2018.

	VOT-metrics				
S^{th}	**EAO ↑**	**Accuracy ↑**	**Robustness ↓**	**# Lost ↓**	**FPS ↑**
0.65	0.276	0.598	0.497	99	25
0.7	0.278	0.601	0.497	99	26
0.75	0.278	0.601	0.497	99	27
0.8	0.278	0.601	0.497	99	27
0.85	0.274	0.601	0.512	102	32
0.9	0.278	0.600	0.512	102	36
0.95	**0.293**	**0.601**	**0.482**	**96**	**41**

Table 4.
Determining the optimal score threshold (S^{th}) for updating the target template under VOT-metrics on VOT2019.

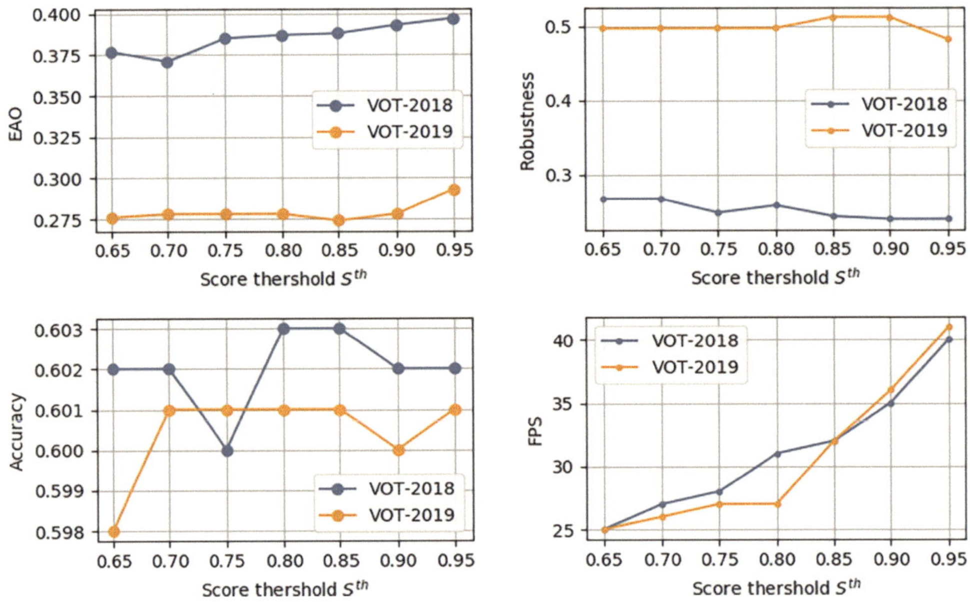

Figure 13.
An illustration on the effect of the tracking performance, with a template update strategy by varying the score threshold S^{th}, NB: This experiments carried out using the checkpoint which include data-augmentation.

occluded by the tree: thus *Update* : *False* flag is generated indicating not to update the target template during such occlusions. This experiment was carried out using $S^{th} >= 0.95$.

Table 5 presents a comparison between no-update SiamMask and incorporating the developed template update strategy: it can be observed that a relative gain of 0.7% and 2.0% in Robustness has been achieved by incorporating template update strategy. Thus, the tracker has encountered less failure than the no-update SiamMask, decreasing from 60 to 58 and 104 to 100 in VOT2018 and VOT2019 benchmarks, respectively. The robustness of the tracker is the crucial element for applications such as automatic robotic cameras where there is no human assistance.

(a) No occlusion, Update template
(b) Full occlusion, Do not update template
(c) Piratical occlusion, Do not update template
(d) No occlusion, Update template

Figure 14.
Visual illustration on how the target template update strategy decides whether to update the template or not based on the similarity score under different occlusion scenario during tracking process, VOT2019 girl dataset. Where blue: Ground truth, red: Tracking result.

(a) No occlusion, Update template
(b) Partial occlusion, Do not update template
(c) Full occlusion, Do not update template
(d) No occlusion, Update template

Figure 15.
Visual illustration on how the target template update strategy decides whether to update the template or not based on the similarity score under different occlusion scenario during tracking process, VOT2019 girl dataset. Where blue: Ground truth, red: Tracking result.

	VOT2018		VOT2019	
	SiamMask	**Ours**	**SiamMask**	**Ours**
EAO ↑	0.380	0.351	0.280	0.268
Accuracy ↑	0.609	0.593	0.610	0.593
Robustness ↓	0.279	**0.272**	0.522	**0.502**
# Lost ↓	60	**58.0**	104	**100.0**
FPS ↑	44	40	44	40

Table 5.
Comparison between no-update SiamMask and incorporating target template update under VOT2018 (left) and VOT2019 (right) benchmarks.

3.7.1 Updating with a previous and a future template

This experiment dedicated to examine the strength and weakness of the" updating with a previous and future frame" template update strategy described in 2.2.1. As can be seen from **Table 6**, the method"updating with previous and a future template" has achieved a relative gain of around 0.7% and 2.5% in Robustness value w.r.t SiamMask in both VOT2018 and VOT2019 benchmark, respectively. This indicates that the"Updating with Previous and Future template" strategy has enhanced the tracker's Robustness, which is the most crucial in automated tracking applications. However, this can not be used for real-time applications as the processing speed is very slow, around 12 FPS on a laptop equipped with NVIDIA GEFORCE GTX1060. The main

	VOT2018		VOT2019	
	SiamMask	**Ours**	**SiamMask**	**Ours**
EAO ↑	0.380	0.357	0.280	0.274
Accuracy ↑	0.609	0.597	0.610	0.597
Robustness ↓	0.279	**0.272**	0.522	**0.497**
# Lost ↓	60	**58**	104	**99**
FPS ↑	**44**	12	**44**	12

Table 6.
Comparison between no-update SiamMask and incorporating target template updating with previous and future template on VOT2018 (left) and VOT2019 (right) benchmark.

computational burden on the tracker is related to the target template feature extraction network. Thus, the tracking algorithm processing speed becomes very slow when the target template is updated with the previous and future template, resulting in a poor FPS.

3.8 Experiment E: Comparison with state-of-the-art trackers

This section compares our tracking framework called VPU_SiamM with other state-of-the-art trackers SiamRPN, SiamMask in the VOT2018 and SiamRPN++, SiamMask in VOT2019.

To take advantage of the incorporated improvements, a tracker named VPU_SiamM has been developed. VPU_SiamM has been trained based on the data augmentation technique incorporating both motion-blur and low-resolution, and during online inference, a target template update strategy is applied.

We have tested our VPU_SiamM tracker on the VOT2018 dataset in comparison with state-of-the-art methods. We compare with the top trackers SiamRPN (winner of the VOT2018 real-time challenge) and SiamMask among the top performer in the VOT2019 challenge. Our tracker obtained a significant relative gain of 1.3% in EAO, compared to the top-ranked trackers. Following the evaluation protocol of VOT2018, we adopt the Expected Average Overlap (EAO), Accuracy (A), and Robustness (R) to compare different trackers. The detailed comparisons are reported in **Table 7**: it can be observed that the VPU_SiamM has achieved top performance on EAO, and R. Especially, our VPU_SiamM tracker outperforms SiamRPN (the VOT2018 real-time challenge winner), achieving a relative gain of 1.3% in EAO and 1.6% in Accuracy and 4% in Robustness. Besides, our tracker yields a relative gain of 4% on Robustness w.r.t

	VOT2018		
Tracker	**EAO↑**	**Accuracy↑**	**Robustness↓**
SiamRPN [21]	0.383	0.586	0.276
SiamMask [11]	0.38	0.609	0.279
VPU_SiamM	**0.393**	0.602	**0.239**

Table 7.
Comparison of our tracker VPU_SiamM with the state-of-the-art trackers SiamRPN and SiamMask in terms of expected average overlap (EAO), accuracy, and robustness (failure rate) on the VOT2018 benchmark.

	VOT2019		
Tracker	**EAO↑**	**Accuracy↑**	**Robustness↓**
SiamRPN++ [14]	0.282	0.598	0.482
SiamMask [11]	0.287	0.594	**0.461**
VPU_SiamM	**0.293**	**0.601**	0.482

Table 8.
Comparison of our tracker VPU_SiamM with the state-of-the-art trackers SiamRPN++ and SiamMask in terms of expected average overlap (EAO), accuracy, and robustness (failure rate) on the VOT2019 benchmark.

both SiamMak and SiamRPN, which is the common vulnerability of the Siamese network-based trackers.

Following previous VOT evaluation, we have evaluated our VPU_SiamM tracker on VOT2019 datasets, which contains 60 challenging testing sequences. As shown in **Table 8**, our VPU_SiamM also achieves the best tracking results on VOT2019 in EAO and Accuracy metrics compared to state-of-the-art trackers SiamMask and SiamRPN++. More specifically, our approach improves the EAO by around 1%.

Submission to VOT-ST 2020 Challenge: Our method (VPU-SiamM) has been submitted to the VOT-ST 2020 challenge [24], and our tracking methods (VPU SiamM) is ranked 16th out of 38 computing tracking methods according to the Expected average overlap (EAO) metrics [24].

4. Conclusions

In this chapter, one of the state-of-the-art tracking algorithms based on siamese networks called SiamMask has been used as a backbone, and two improvements have been affixed, each addressing different aspects of the tracking task.

The developed data augmentation technique for incorporating low-resolution and motion-blur has been evaluated separately and jointly, achieving state-of-the-art results in the VOT2018 and VOT2019 benchmarks. From the evaluation results, it is clear to conclude that the data augmentation technique has played an essential role in improving the overall performance of the tracking algorithm. It has outperformed the SiamMask results in both VOT2018 and VOT2019 benchmarks. In contrast, among the three data augmentation techniques, the data augmentation technique for incorporating both motion-blur and low-resolution outperforms the rest in terms of EAO in VOT2018 and VOT2019 benchmarks. Nevertheless, the data-augmentation for incorporating only motion-blur has achieved a top performance according to the Accuracy metric in both VOT2018 and VOT2019 benchmarks. However, the Accuracy is less significant as it only considers the IOU during a successful tracking. According to the VOT ranking method, the EAO value is used to rank tracking methods. Therefore the data augmentation technique for incorporating both motion-blur and low-resolution is ranked top among the others. This indicates that the data-augmentation technique has contributed to the improvement of the overall tracker performance.

Comparable results on VOT2018 and VOT2019 benchmarks confirm that the robust target template update strategy that utilizes both the initial ground truth template and a supplementary updatable template and avoiding template updates during severe occlusion can significantly improve the tracker's performance with respect to SiamMask results while running at 41 FPS.

A tracker named VPU_SiamM was trained based on the presented approach, and it was ranked 16th out of 38 submitted tracking methods in the VOT-ST 2020 challenge [24].

Acknowledgements

This work has been partially supported by the Spanish Government through its TEC2017-88169-R MobiNetVideo project.

Author details

Awet Haileslassie Gebrehiwot[1*], Jesus Bescos[2] and Alvaro Garcia-Martin[2]

1 Czech Technical University in Prague, Prague, Czech Republic

2 Universidad Autonoma de Madrid, Madrid, Spain

*Address all correspondence to: awethaileslassie21@gmail.com

References

[1] Tian B, Yao Q, Gu Y, Wang K, Li Y. Video processing techniques for traffic flow monitoring: A survey. In: 14th International IEEE Conference on Intelligent Transportation Systems, ITSC 2011, Washington, DC, USA, October 5-7, 2011. IEEE; 2011. pp. 1103-1108

[2] Zeng M, Guo G, Tang Q. Vehicle human-machine interaction interface evaluation method based on eye movement and finger tracking technology. In: HCI International 2019 - Late Breaking Papers - 21st HCI International Conference, HCII 2019, Orlando, FL, USA, July 26-31, 2019, Proceedings (C. Stephanidis, ed.), vol. 11786 of Lecture Notes in Computer Science. Springer; 2019. pp. 101-115

[3] Brandes S, Mokhtari Z, Essig F, Hünniger K, Kurzai O, Figge MT. Automated segmentation and tracking of non-rigid objects in time-lapse microscopy videos of polymorphonuclear neutrophils. Medical Image Anal. 2015;**20**(1):34-51

[4] Nägeli T, Alonso-Mora J, Domahidi A, Rus D, Hilliges O. Real-time motion planning for aerial videography with real-time with dynamic obstacle avoidance and viewpoint optimization. IEEE Robotics Autom. Lett. 2017;**2**(3):1696-1703

[5] Esterle L, Lewis PR, McBride R, Yao X. The future of camera networks: Staying smart in a chaotic world. In: Arias-Estrada MO, Micheloni C, Aghajan HK, Camps OI, Brea VM, editors. Proceedings of the 11th International Conference on Distributed Smart Cameras, Stanford, CA, USA, September 5-7, 2017. ACM; 2017. pp. 163-168

[6] Chen YF, Everett M, Liu M, How JP. Socially aware motion planning with deep reinforcement learning. CoRR. 2017;**abs/1703.08862**

[7] Aggarwal JK, Xia L. Human activity recognition from 3d data: A review. Pattern Recognition Letters. 2014;**48**: 70-80

[8] Bertinetto L, Valmadre J, Henriques JF, Vedaldi A, Torr PHS. Fully-convolutional siamese networks for object tracking. In: Computer Vision - ECCV 2016 Workshops - Amsterdam, The Netherlands, October 8-10 and 15-16, 2016, Proceedings, Part II (G. Hua and H. Jégou, eds.), vol. 9914 of Lecture Notes in Computer Science. 2016. pp. 850-865

[9] Matej Kristan EA, Matas J. The visual object tracking VOT2017 challenge results. In: 2017 IEEE International Conference on Computer Vision Workshops, ICCV Workshops 2017, Venice, Italy, October 22-29, 2017. IEEE Computer Society; 2017. pp. 1949-1972

[10] Russakovsky O, Deng J, Su H, Krause J, Satheesh S, Ma S, et al. Imagenet large scale visual recognition challenge. International Journal of Computer Vision. 2015;**115**(3):211-252

[11] Wang Q, Zhang L, Bertinetto L, Hu W, Torr PHS. Fast online object tracking and segmentation: A unifying approach. In: IEEE Conference on Computer Vision and Pattern Recognition, CVPR 2019, Long Beach, CA, USA, June 16-20, 2019. Computer Vision Foundation / IEEE; 2019. pp. 1328-1338

[12] Zhou J, Wang P and Sun H. Discriminative and Robust Online Learning for Siamese Visual Tracking. 2019

[13] Zhang Z, Peng H, Wang Q. Deeper and wider siamese networks for real-

time visual tracking. CoRR. 2019;**abs/1901.01660**

[14] Li B, Wu W, Wang Q, Zhang F, Xing J, Yan J. Siamrpn++: Evolution of siamese visual tracking with very deep networks. In: arXiv preprint arXiv: 1812.11703. 2018

[15] Wang G, Luo C, Xiong Z, Zeng W. Spm-tracker: Series-parallel matching for real-time visual object tracking. CoRR. 2019;**abs/1904.04452**

[16] Matej Kristan EA. The seventh visual object tracking VOT2019 challenge results. In: 2019 IEEE/CVF International Conference on Computer Vision Workshops, ICCV Workshops 2019, Seoul, Korea (South), October 27-28, 2019. IEEE; 2019. pp. 2206-2241

[17] Shorten C, Khoshgoftaar TM. A survey on image data augmentation for deep learning. J. Big Data. 2019;**6**:60

[18] Wang Z, Chang S, Yang Y, Liu D, Huang TS. Studying very low resolution recognition using deep networks. In: 2016 IEEE Conference on Computer Vision and Pattern Recognition, CVPR 2016, Las Vegas, NV, USA, June 27-30, 2016. IEEE Computer Society; 2016. pp. 4792-4800

[19] Wang Q, Teng Z, Xing J, Gao J, Hu W, Maybank SJ. Learning attentions: Residual attentional siamese network for high performance online visual tracking. In: 2018 IEEE Conference on Computer Vision and Pattern Recognition, CVPR 2018, Salt Lake City, UT, USA, June 18-22, 2018. IEEE Computer Society; 2018. pp. 4854-4863

[20] Zhu Z, Wang Q, Li B, Wu W, Yan J, Hu W. Distractor-aware siamese networks for visual object tracking. In: Computer Vision - ECCV 2018 - 15th European Conference, Munich, Germany, September 8-14, 2018, Proceedings, Part IX (V. Ferrari, M. Hebert, C. Sminchisescu, and Y. Weiss, eds.), vol. 11213 of Lecture Notes in Computer Science. Springer; 2018. pp. 103-119

[21] Li B, Yan J, Wu W, Zhu Z, Hu X. High performance visual tracking with siamese region proposal network. In: 2018 IEEE Conference on Computer Vision and Pattern Recognition, CVPR 2018, Salt Lake City, UT, USA, June 18-22, 2018. IEEE Computer Society; 2018. pp. 8971-8980

[22] Zhang L, Gonzalez-Garcia A, van de Weijer J, Danelljan M, Khan FS. Learning the model update for siamese trackers. In: 2019 IEEE/CVF International Conference on Computer Vision, ICCV 2019, Seoul, Korea (South), October 27 - November 2, 2019. IEEE; 2019. pp. 4009-4018

[23] Kristan EAM. The visual object tracking vot2013 challenge results. In: 2013 IEEE International Conference on Computer Vision Workshops. 2013. pp. 98-111

[24] Kristan M, Leonardis A, Matas J, Felsberg M, Pflugfelder R, Kämäräinen J-K, et al. "The eighth visual object tracking vot2020 challenge results," in Computer Vision – ECCV 2020 Workshops (A. Bartoli and A. Fusiello, eds.), (Cham), Springer: International Publishing; 2020. pp. 547-601

Chapter 4

Cognitive Visual Tracking of Hand Gestures in Real-Time RGB Videos

Richa Golash and Yogendra Kumar Jain

Abstract

Real-time visual hand tracking is quite different from commonly tracked objects in RGB videos. Because the hand is a biological object and hence suffers from both physical and behavioral variations during its movement. Furthermore, the hand acquires a very small area in the image frame, and due to its erratic pattern of movement, the quality of images in the video is affected considerably, if recorded from a simple RGB camera. In this chapter, we propose a hybrid framework to track the hand movement in RGB video sequences. The framework integrates the unique features of the Faster Region-based Convolutional Neural Network (Faster R-CNN) built on Residual Network and Scale-Invariant Feature Transform (SIFT) algorithm. This combination is enriched with the discriminative learning power of deep neural networks and the fast detection capability of hand-crafted features SIFT. Thus, our method online adapts the variations occurring in real-time hand movement and exhibits high efficiency in cognitive recognition of hand trajectory. The empirical results shown in the chapter demonstrate that the approach can withstand the intrinsic as well as extrinsic challenges associated with visual tracking of hand gestures in RGB videos.

Keywords: hand tracking, faster R-CNN, hand detection, feature extraction, scale invariant, artificial neural network

1. Introduction

Hand Gestures play very significant roles in our day-to-day communication, and often they convey more than words. As technology and information are growing rapidly in every sector of our life, interaction with machines has become an unavoidable part of life. Thus, a deep urge for natural interaction with machines is growing all around [1, 2]. One of the biggest accomplishments in the domain of Hand Gesture Recognition (HGR) is Sign language recognition (SLR) where machines interpret the static hand posture of a human standing in front of a camera [3]. Recently, implementation of HGR-based automotive interface in BMW cars is very much appreciated. Here, five gestures are used for contactless control of music volume and incoming calls while driving [4]. Project Soli is the ongoing project of Google's Advanced Technology; in this project a miniature radar is developed that understands the real-time motion of the human hand at various scales [5].

Hand gestures are very versatile as they comprise static as well as dynamic characteristics, physical as well as behavioral characteristics, for example, movement in any direction, fingers can bend to many angles. Hand skeleton has a complex structure with a very high freedom factor, and thus its two-dimensional RGB data sequence has unpredictable variations. Visual recognition of dynamic hand gestures is complex because the complete process requires the determination of hand posture along with a cognitive estimation of the trajectory of motion of that posture [3, 6–9]. Due to these intricacies to date, vision-based HGR applications mainly dominate with static hand gesture recognition.

2. Challenges in online tracking hand motion

In context with computer vision and pattern recognition, a human hand is described as a biological target with a complex structure. Uneven surface, broken contours, and erratic pattern of movement are some of the natural characteristics that complicate DHGR [10]. Thus, in comparison to the other commonly tracked moving object, a hand is a non-rigid subtle object and covers a very small area in the image frame. The scientific challenges accompanied in the online tracking of the hand region in an unconstrained environment in RGB images captured using a simple camera are categorized as follows: [3, 4, 6–11].

i. Intrinsic Challenges: Intrinsic challenges are related to a target that is "Hand" physical and behavioral nature. The features such as

   Hand Appearance: The number of joints in the hand skeleton, the appearance of the same hand posture has a large variation, known as shape deformation. Different postures have a wide difference in occupancy area in an image frame, and some postures only cover 10% of the image frame, which is a very small target size in computer vision. In a real-time unconstrained environment, the two-dimensional (2-D) posture shows large variation during movement.

   Manner of Movement: There is a large diversity among human beings in performing the gesture of the same meaning, in terms of speed and path of movement. The moving pattern of the hand is erratic, irregular, and produces blur in the image sequence. Furthermore, the two-dimensional data sequence of a moving hand is greatly affected by background conditions, thus tracking and interpretation of dynamic hand gestures are a challenging task in the HGR domain. The unpredictable variation in target trajectory makes the detection and classification process complex in pattern recognition.

ii. Extrinsic Challenges: These challenges mainly arise due to the environment in which the hand movement is captured. Some of the major factors that deeply impact the real-time visual tracking of the dynamic hand gestures are as follows:

   Background: In the real-time HGR applications, backgrounds are unconstrained, we cannot use fixed background models to differentiate between the foreground and the background. Thus, the core challenge in the design of a real-time hand tracking system is the estimation of discriminative features between background and target hand posture.

   Illumination: Illumination conditions in real-time applications are uneven and also unstable. Thus, 2-D (two-dimensional) projection of the 3-D (three-dimensional) hand movement produces loss of information in RGB images. This loss is the major reason for errors in the visual tracking of hand movement.

Presence of other skin color objects in the surroundings: The presence of objects with similar RGB values such as the face, neck, arm, etc., is the serious cause for track loss in the RGB-based visual tracking techniques.

3. Components of DHGR

There are four main components in cognitive recognition of dynamic hand gestures [3, 10–12].

i. Data Acquisition.

ii. Interest Region Detection.

iii. Tracking of Interest Region.

iv. Classification of Trajectory.

In Dynamic Hand Gesture Recognition (DHGR), acquisition of signals plays a very important role in deciding the technique to recognize and deduce the hand pattern into meaningful information. Contact-based sensors and contactless sensors are two main types of sensors to acquire hand movement signals. Contact-based sensors are those sensors that are attached to the body parts of a user example. Data gloves are hand gloves, accelerometers are attached the arm region, and egocentric sensors are put on the head to record hand movement. Wearable sensor devices are equipped with inertial, magnetic sensors, mechanical, ultrasonic, or barometric [7]. Andrea Bandini et al. [13], in their survey, presented many advantages of egocentric vision-based techniques as they can acquire hand signals very closely. Although the contact-based techniques require fewer computations, but wearing these devices gives uneasiness to the subject. Due to the electrical and magnetic emission of signals, it is likely to produce hazardous effects on the human body.

Contactless sensors or vision-based sensor technology is becoming encouraging technology to develop natural human-machine interfaces [1–4, 14]. These devices consist of visual sensors, with a single or a group of cameras situated at a distance from the user to record the hand movement. In vision-based methods, the acquired data is image type, a user does not have to wear any devices, and he can move his hand naturally in an unconstrained pattern. The important assets of vision-based techniques are large flexibility for users, low hardware requirements, and no health issues. These methods have the potential to develop any natural interface for remote human-machine interaction, this can ease the living of physically challenged or elderly people with impaired mobility [2, 9, 15].

In vision-based methods, the information is two-dimensional, three-dimensional, or multiview images. Two-dimensional images are RGB images with only intensity information about the object, captured using simple cameras and. Three-dimensional images are captured from advanced sensor cameras such as Kinect, Leap Motion, Time of flight, etc.; these cameras collect RGB along with depth information of the object in the scene. The third and the most popular choice in HGR is multiview images; here two or more cameras are placed at different angles to capture the hand movement from many views [3, 6, 8].

Wang J. et al. [16] used two calibrated cameras to record hand gestures under stable lighting conditions. They initially segmented the hand region using YCbCr color space and then applied SIFT algorithm for feature extraction. After then, they tracked using Kalman Filter. But due to similarity with other objects, the author imposes position constraints to avoid track loss.

Poon G. et al. [17] also supported multiple camera setups that can observe the hand region from diversified angles to minimize the errors due to self-occlusion. They proposed three camera setups to recognize bimanual gestures in HGR. Similarly, Bautista A.G. et al. [18] used three cameras in their system to avoid complex background and illumination. Marin G. et al. [19] suggested combining Kinect data with Leap motion camera data to exploit the complementary characteristics of both the cameras. Kainz O. et al. [20] combined leap motion sensor signals and surface electromyography signals to propose a hand tracking scheme.

Andreas Aristidou discussed that high complexity in hand structure and movement make the animation of a hand model a challenge. They preferred a marker-based optical motion capture system to acquire the orientation of the hand [21]. With the same opinion, Lizy Abraham et al. [22] placed infrared LEDs on the hand to improve the consistency of accuracy in tracking. According to the study conducted by Mais Yasen et al. [9], surface electromyography (sEMG) as wearable sensors and Artificial Neural Network (ANN) as classifiers are the most preferable choices in hand gesture recognition.

The important factor in HGR is that information obtained using a monocular camera is not sufficient to extract the moving hand region. The loss of information in RGB images is maximum due to unpredictable background, self-occlusion, illumination variation, and erratic pattern of the hand movement [8, 10, 14].

The second component in the design of DHGR is description of the region of interest or "target modeling." In this section, features that are repetitive, unique, and invariant to general variations, e.g., illumination, rotation, translation of the hand region are collected. These features model the target of tracking and are responsible for detecting and localizing the target in all frames of a video. This step is very significant because it helps to detect the target in an unconstrained environment [10, 12].

Li X. et al. [12] presented a very detailed study of the building blocks of visual object tracking and the associated challenges. They stated that effective modeling of the appearance of the target is the core issue for the success of a visual tracker. Practically, effective modeling is greatly affected by many factors such as target speed, illumination conditions, state of occlusion, complexity in shape, and camera stability, etc. Skin color features are the most straightforward characteristic of the hand used in the HGR domain to identify the hand region in the scene. Huang H. et al. simply detected skin color for contour extraction and then classified them using VGGNet [23]. M. H. Yao et al. [24] extracted 500 particles using the CAMShift algorithm for tracking the moving hand region. In this case, the real-time performance of the HGR system decreases when a similar color object (face or arm region) interferes. As the number of particles increases the complexity of the system increases. The HGR technique proposed by Khaled H. et al. [25] emphasized the use of both shape and skin color features for hand area detection because of background conditions, shadows, visual overlapping of the objects. They stated that noise added due to camera movement is one of the major problems in real-time hand tracking. Liu P. et al. [26] proposed a single-shot multibox detector ConvNet architecture that is like Faster R-CNN to detect hand gestures in a complex environment. Bao P. et al. [27] expressed that since the size of hand posture is very small, therefore misleading behavior or the overfitting problem becomes prominent in regular CNN.

In the method discussed in [10], we have shown that though the local representation of the hand is a comparatively more robust approach to detect the hand region, but they often suffer from background disturbance in a real-time tracking. In general, hand-crafted features result in large computations and loss of trajectory visual while tracking in real-time hand movement is very common. Henceforth, it is difficult for hand-crafted features to perfectly describe all variations in target as well as background [10, 12]. According to Shin J. et al. [28], the trackers that visually trace the object, based on appearance and position, must have a high tolerance for appearance and position. Tran D. et al. [29] initially detected the palm region from depth data collected by Kinect V2 skeletal tracker followed by morphological processing. They determined hand contour using a border tracing algorithm on binary image converted using a fixed threshold. After detecting fingertip by K-cosine algorithm, hand posture is classified using 3DCNN.

Matching of hand gesture trajectory is another important phase in the cognitive recognition of DHGR. The main constrain in generating similarity index in HGR is the speed of hand motion and the path of movement. Both these factors are highly dependent on the user's mood and surrounding conditions at the instant of movement. Similarity matching based on distance metrics generally fails to track efficiently as hand gestures of the same meaning do not follow the same path always.

Dan Zhao et al. [30] used a hand shape fisher vector to find the movement of the finger and then classified it by linear SVM. Plouffe et al. [31] proposed Dynamic Time Wrapping (DTW) to match the similarity between target and trained gesture. In [32], a two-level speed normalization procedure is proposed using DTW and Euclidean distance-based techniques. In this method, for each test gesture, 10 best-trained gestures are selected using the DTW algorithm. Out of these 10 gestures, the most accurate gesture is selected by calculating Euclidean distance. Pablo B. et al. [33] suggested a combination of the Hidden Markov Model (HMM) and DTW, in the prediction stage.

4. Proposed methodology

The proposed system is designed by using a web camera; it is a simple RGB camera. The use of the RGB camera is limited in the field of hand gesture tracking because of various difficulties as discussed above (**Figure 1**).

The proposed system is divided into three modules:

4.1 Module I

This module is also known as the "hand detection module." Here the posture of the hand, which is used by the user in real-time hand movement events, is detected. When the user moves his hand in front of the web camera attached to any machine acquires a video of 5–6 seconds at a rate of 15 frames per second. This video comprises a raw data sequence of length 100–150 frames; it is saved in a temporary folder, resizing all the frames to size [240, 240]. In this module, detection of an online Active

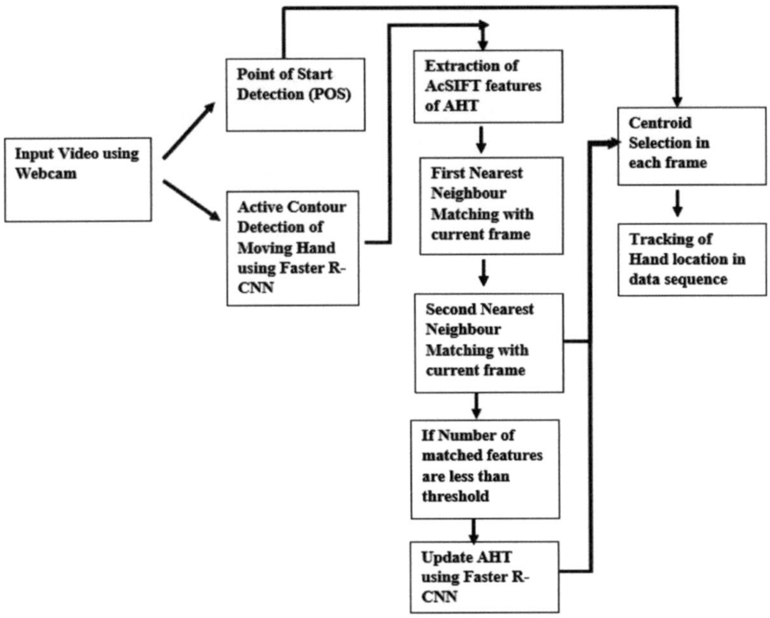

Figure 1.
Architecture of the proposed system.

Hand Template (AHT) is made using Faster Region-based Convolutional Neural Network (Faster R-CNN).

4.1.1 Faster R-CNN

We have proposed the design of an online hand detection scheme (AHT) using Faster Region-based Convolutional Neural Network (Faster R-CNN) [34], on Residual Network (ResNet101) [35], a deep neural architecture. Three major issues that are encountered in online tracking of hand motion captured using simple cameras are as follows:

1. A hand is a versatile object in comparison with other objects. The area occupied in the image frame has a high variation that depends on the posture selected.

2. It is not fixed that the subject starts the motion from the first frame or the fixed position in the frame.

3. Anthropometric and scale variation in the hand are very prominently seen during hand movement in RGB images.

Thus, the essential requisite of any technique is to cope with the abovementioned factors. In the proposed method, these issues are solved by using Faster-RCNN, a Deep Neural Network (DNN) architecture. Deep learning algorithms (DLAs) are models for a machine to learn and execute any task as human beings perform. Deep networks directly learn features from raw data by exploiting local information of the target, with no manual extraction or elimination of background. Convolutional Neural

Network (ConvNet) is a powerful tool in the computer vision field that mainly deals with images.

Ren S. et al. [34] modified fast RCNN to Faster Region-based Convolutional Neural Network (Faster R-CNN). They added a region proposal network (RPN) (a separate CNN network) that simultaneously estimates objectness score and regresses the boundaries of the object using the anchor box concept.

The architecture of the proposed Faster R-CNN developed on ResNet 101 is shown in **Figure 2**. Region Proposal Network (RPN) is an independent small-sized ConvNet, designed to directly generate region proposals from an image of any size without using a fixed edge box algorithm. The process of RPN is shown in **Figure 3**; here region proposals are generated from the activation feature map of the last shared

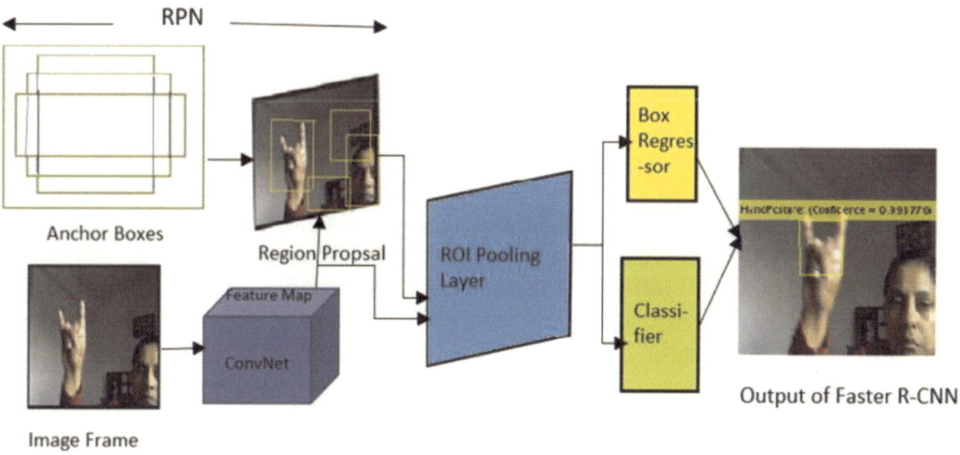

Figure 2.
The architecture of the proposed faster R-CNN.

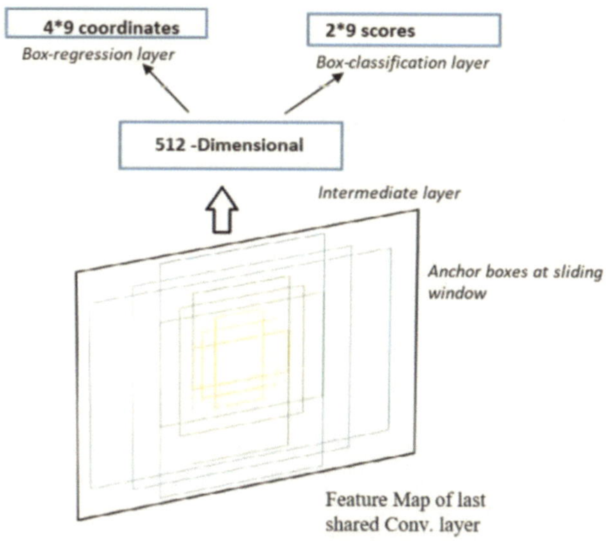

Figure 3.
Process in RPN.

convolutional layer between the RPN network and Fast-RCNN. It is implemented with an $m \times m$ convolutional layer followed by two siblings: box regression layer and box classification layer each of size 1×1. At each sliding grid, multiple regions are proposed depending upon the number of anchor boxes (Q). Each predicted region is classified by a score and four tuples (x, y, L, B) where (x, y) are the coordinates of the top left corner of the bounding box, L and B are the length and breadth of the box. If $M \times N$ is the size of the feature map and number of anchors, the technique is Q, then total anchors created will be $M \times N \times Q$.

Anchor boxes are bounding boxes with predefined height and width to capture the scale and aspect ratio of the target object. There are pyramids of anchors. The anchor-based method is translation invariant and detects objects of multiple scales and aspect ratios. For every tiled anchor box, the RPN predicts the probability of object, background, and intersection over union (IoU) values. The advantage of using the anchor boxes in a sliding window-based detector is to detect, encode, and classify the object in the region in a single process [34].

The design of the proposed Faster-RCNN technique is accomplished on Residual network (Resnet) resnet 101. Resnet architecture network was proposed in 2015 by Kaiming He et al. [35], to ease the learning process in a deeper network. They exhibited that a resnet architecture eight times deeper than VGG16 still has less complexity in training on ImageNet dataset. The proposed use of resnet 101 in the design of Faster R-CNN solves the complex problems in object classification by using a large number of hidden layers without increasing the training error. Furthermore, the network does not have a vanishing and exploding gradient problem because of the "skip connection" approach.

4.2 Module II

This module handles the feature extraction process of the AHT that helps in the continuous localization of the moving hand region. Our method processes a hybrid framework that combines Scale Invariant Feature Transform (SIFT) and Faster-RCNN. A framework with hybrid characteristics is selected because in real-time movement, the geometrical shape of any posture changes many times, and thus it is difficult to detect the moving hand region with only hand-crafted features i.e., SIFT. Whenever the posture is changed above the threshold (number of matched features <= 3), then AHT is determined using Faster R-CNN, and the previous AHT is updated with new AHT. During this process, a bounding box is also constructed around the centroid of the hand movement to determine the current two-dimensional area covered by the hand region.

4.2.1 Scale invariant feature transform (SIFT)

In motion modeling, we have used SIFT algorithm designed by David Lowe [36], for local feature extraction of AHT. As compared with global features such as color, contour, texture, local features have high distinctiveness, better detection accuracy toward local image distortions, viewpoint change, and partial occlusion. Therefore, SIFT detects the object in the cluttered background without performing any segmentation or preprocessing algorithms [36, 37]. The combination of SIFT and Faster-RCNN is helpful in real-time fast-tracking of the non-rigid subtle object hand.

SIFT algorithm comprises of feature detector as well as a feature descriptor. In general, features are high-contrast areas example point, edge, or small image patch, in an image. These features are extracted such that they are detectable even in noise, scale variation, and during the change in illumination. Each SIFT feature is defined by four parameters: $f_i = \{p_i, \sigma_i, \varphi_i\}$, where $p_i = (x_i, y_i)$ is the 2D position of SIFT keypoint, σ_i is the scale, φ_i is gradient orientation within the region. Each key point i d is described by 128-dimensional descriptor d [36].

In our approach, we find the SIFT features of the AHT template obtained in module-I, since it contains only the target hand posture and is small as compared with the image frame [240, 240]. Therefore, this approach saves time in matching unnecessary features and pruning them further [20, 21].

Let there be m key features in AHT frame, given as $S_{AHT} = \{f_i\}^m$, where f_i is the feature vector at i^{th} location. Let $S_{cur} = \{f_j\}^k$ are k numbers of SIFT features in the current frame, where f_j is the SIFT feature at j^{th} location. We use the best-bin-first search method that identifies the nearest neighbors of AHT features with current frame features. The process of SIFT target recognition and localization in the subsequent frames of a video is accomplished in three steps:

Initially, we find the first nearest neighbors (FNN) of all the SIFT features in AHT with SIFT features in the current frame. The First Nearest Neighbors (FNNs) are defined as the pairs of key points in two different frames with a minimum sum of squared differences for the given descriptor vector

$$distanceFNND(a_{AHT}, b_{cur}) = \sqrt{\sum_{i=1}^{128} (a_i - b_i)} \tag{1}$$

where a_{AHT} and b_{cur} are descriptor vectors of features in AHT and current frame, respectively.

In the second step, matching is improved by performing Lowe's Second Nearest Neighbor (SNN) test using Eq. (2).

$$\frac{distance(a_{AHT}, b_{cur})}{distance\ (a_{AHT}, c_{cur})} > 0.8 \tag{2}$$

SNN test is done by calculating the ratio between the FNND of a_{AHT} feature with two nearest neighbors b_{cur} and c_{cur} in the current frame.

Further to find the geometrically consistent points, we apply the geometric verification test (Eq. (3)) on the key points obtained after SNN.

$$\begin{bmatrix} x^* \\ y^* \end{bmatrix} = vR(\alpha)\begin{bmatrix} x \\ y \end{bmatrix} + \begin{bmatrix} T_x \\ T_y \end{bmatrix} \tag{3}$$

Here v is isotropic scaling, α is rotation parameter, (T_x, T_y) are translation vectors for the ith SIFT keypoint located at a distance (x, y).

4.3 Module III

This module deals with the cognitive recognition of the trajectory. Here the cognitive recognition means vision-based intellectual development of machine for the interpretation of hand movement. Because hand movements do not have a fixed pattern, by nature movement patterns are erratic. Due to this characteristic, till now static hand gesture recognition is more preferred than dynamic hand gesture recognition. We have determined the centroids of hand location in the tracked frames. To derive the meaning of hand movement, we have used the modified back-propagation Artificial Neural Network (m-BP-ANN) Match of test trajectory to train database. This cognitive stage is very significant for DHG because the way we collect and transform the centroid of hand movement C_{HM} of every frame in a particular data sequence, helps to classify the hand gesture. In the proposed system we have kept this stage simple but efficient because complex algorithms increase the error rate and time of interpretation.

We have made use of the concept of the quadrant system of the Cartesian plane to transform the image frame into a 2-D plane. The two-dimensional Cartesian system divides the plane of the frame into four equal regions called Quadrants. Each quadrant is bound by two half-axes, with the center in the middle of a frame. The translation of the image frame axis to a Cartesian axis is done using Eqs. (4) and (5):

$$x_c = (C_{HMx} - I_x)/n_x \tag{4}$$

$$y_c = (C_{HMy} - I_y)/n_y \tag{5}$$

Here I_x, I_y are the dimensions of the image frame [240, 240] and n_x, n_y [12, 12] are normalization factors for the X and Y-axis. To convert the hand trajectory into meaningful command, we have applied Modified Back-Propagation of Artificial Neural Network (mBP-ANN) using start and end location of the hand gesture.

Back-propagation (BP) is a supervised training procedure in feed-forward neural networks. It works on minimizing the cost function of the network using the delta rule or gradient descent method. The value of the weights with which we obtain the minimum cost function is the solution for the given learning problem. The error function $'E_f''$ is defined as the mean square sum of the difference between the actual output value of the network (a_j) and the desired target value (t_j) for the jth neuron. E_f calculated for N_L number of output neurons in$''L''$ a number of layers are given as Eq. (6):

$$E_f = 1/2\sum_{p=1}^{P}\sum_{j=1}^{N_L}(t_j - a_j)^2 \tag{6}$$

The minimization of the error function is carried out using gradient descent or delta rule. It determines the amount of weight update based on gradient direction along with step size. It is given by Eq. (7):

$$\frac{\partial C(t+1)}{\partial \delta_{ij}(t)} = \frac{\partial C(t+1)}{\partial w_{ij}(t+1)} \text{x} \frac{\partial w_{ij}(t+1)}{\partial \delta_{ij}(t)} \tag{7}$$

In the traditional BP, the optimization of the multidimensional cost function is difficult because step size is fixed, since the performance parameters are highly dependent on the learning rate δ. Hence, to overcome the problems of fixed step size and slow learning, we use adaptive learning and momentum term to modify BP. The updated weight value at any node is given by Eq. (8):

$$\Delta w_{ij}(t) = \eta\delta_j a_i + m\Delta w_{ij}(t-1) \tag{8}$$

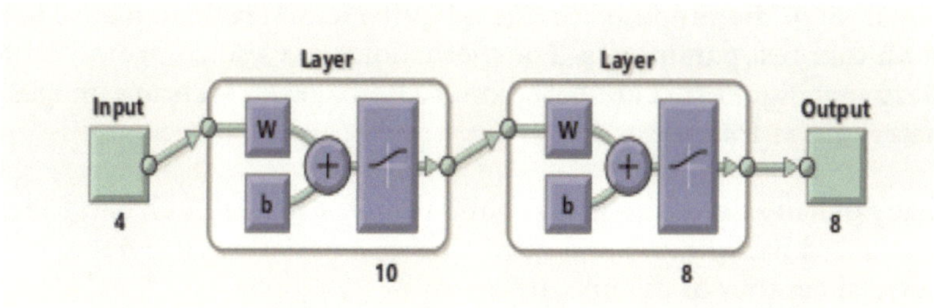

Figure 4.
Architecture of the proposed ANN model.

The term momentum ($0 < m < 1$) updates the value of weight using the previous value of it. Adaptive learning rates help to learn the characteristic of the cost function. If the effort function is decreasing, then the learning rate will increase, and vice versa [38].

In the proposed prototype, we have developed eight vision-based commands to operate and machine remotely by showing hand gestures. The proposed model of ANN has three layers, input layer, hidden layer, and output layer as shown in **Figure 4**. The input layer has 4 neurons, the hidden layer has 10 neurons, and the outer layer consists of 8 neurons.

5. Experimental analysis

In this research work, we have taken three hand postures (as shown in **Table 1**) to demonstrate the vision-based tracking efficiency of our proposed concept. It is the unique feature of this work as most of the techniques demonstrate tracking of hand movements performed by a single posture [32]. For consolidated evaluation, we have taken approximately 100 data sequences captured in different environments as shown in **Figure 5**. Our database is a collection of publicly available dataset [32] and self-prepared data sequence. In [32], hand movements are mainly performed by a single hand posture (Posture III as shown in **Table 1**) and in a constrained laboratory environment.

In self-prepared dataset, we have collected hand movements performed by six participants of three different age groups: two kids (age 10–16 years), two adults (age 20–40 years), and two seniors (age 65 years). In this, the hand movement is carried out using three different postures (as illustrated in **Table 1**), in linear as well as circular pattern. In self-collected dataset, 15 frames per second are taken through the web camera, and gesture length varies from 120 to 160 frames.

		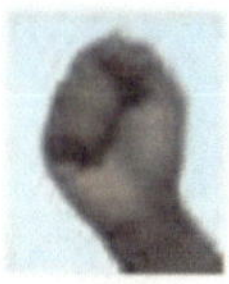
Posture I	Posture II	Posture III

Table1.
Types of postures used in the proposed system.

The evaluation of the proposed online adaptive hand tracking methodology is carried out on four test parameters. The methodology is also compared with the contemporary techniques that are based on RGB images or webcam images. The four test parameters are as follows:

1. Accuracy in hand detection in real-time complex images, i.e., video is captured in unconstrained background and covers natural variations occurring in geometrical contour of the postures.

2. Parametric evaluation of the proposed Faster R-CNN on resnet101 architecture on training and validation data.

3. The efficiency of a hybrid tracking system in complex environment.

4. Effectiveness of cognitive recognition of hand trajectory as machine command.

5.1 Accuracy in hand detection

Figure 6 demonstrates the outcome of the hand recognition stage of different data sequences captured (using three hand postures demonstrated in **Table 1**) in different backgrounds under d

ifferent illumination conditions. To test the accuracy of the hand detection scheme in recognizing the hand region, we have considered nearly all possible combinations:

Figure 5.
Dataset for training faster R-CNN.

only the hand is visible in the camera view, subject face along with arm region is in the camera view, illumination conditions are unstable, background has same color as the hand region, etc. Thus, the hand detection results in **Figure 6** illustrate the following distinguishing key features of our proposed system:

i. Large diversity is present in hand shape and sizes also, the same posture differs in geometrical shapes and area of coverage in the image frame. Our technique does not require any foreground and background modeling. It detects the hand region by automatic learning, the discriminative deep features of the hand postures.

ii. The subject's state of mind at the instance of hand movement is not alike. Thus, it is not necessary that the hand is completely visible from the first frame. Our technique is not affected by the location from which the user starts their motion, it is also unaffected by the face region or other body parts of the user present in the data sequence.

5.2 Parametric evaluation of module i

The proposed hand detection module, developed on Faster R-CNN architecture, has been evaluated on the following parameters:

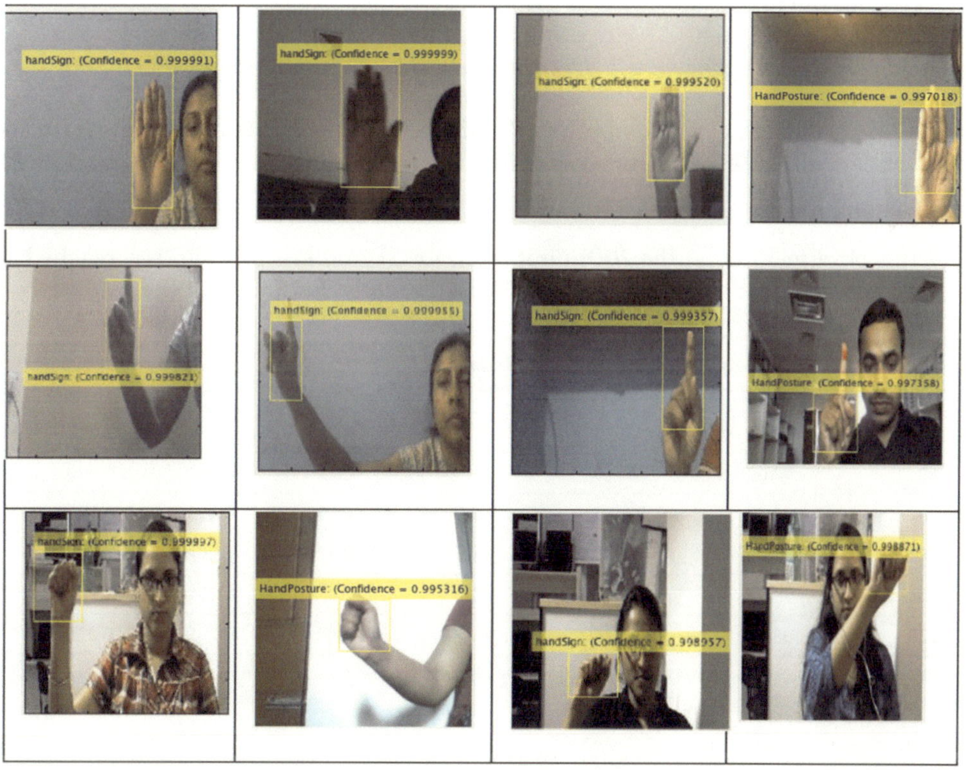

Figure 6.
Various outcomes of module-I (simple background, complex background, the subject is also visible in camera range).

i. Accuracy: It is the parameters by which the network is evaluated and selected. It gives the count of accurate predictions.

ii. Loss: The loss curve is the most useful diagnostic curve that accounts for variation in the predicted and actual value. Loss information helps to learn the optimization behavior of the model parameters.

iii. Model Behavior: It talks about the learning behavior of the model. Learning pattern helps to diagnose the character of the train or validation dataset concerning the problem domain.

iv. Root Mean Square Error (RMSE): It calculates the standard deviation between the actual value and the predicted value. The RMSE is applied in regression analysis and classification of the predicted bounding box with the ground truth bounding box. It is calculated during the training process for both train data and validation data.

Table 2 illustrates detailed performance outcomes of the proposed Faster R-CNN based on the abovementioned parameters. The observations are taken at intervals of 50, 100, 150, 200, 220 iterations. The outcomes illustrate following points of our proposed architecture on Faster R-CNN constructed on resnet101:

1. As the number of iterations increases, the accuracy of train data increases, and it reaches the maximum value at the 200th iteration.

2. Validation data achieve the maximum accuracy at 220th iterations.

3. There is a linear decrement in the RMSE and the loss values of both the train and validation dataset. This linear decrement reflects the stable learning behavior of the proposed model.

4. It is observed that at the 200th iteration, RMSE and loss of train data reached at its minimum value of 0.14 and 0.154, respectively. Similarly, in the case of the

No. of iteration	Train data accuracy	Validation data accuracy	Train data RMSE	Validation data RMSE	Train data loss	Validation data loss
1	30.24	78.26	0.23	0.22	2.9331	2.2522
50	97.07	98.80	0.19	0.19	0.9396	0.6902
100	98.32	98.87	0.15	0.19	0.4791	0.6049
150	99.03	98.85	0.16	0.17	0.2671	0.5956
200	99.07	97.86	0.14	0.17	0.1544	0.55544
220	98.92	98.76	0.17	0.17	0.2052	0.6262

Based on above outcomes, the characteristic features of the proposed trained resnet101 are:
Accuracy: 98.76%
Loss: 0.17
The behavior of the Network: Well fit.

Table 2.
Outcomes in the training process of the proposed faster R-CNN model.

validation dataset, the value of RMSE and loss reached their minimal at the 200th iteration.

5.3 Efficiency of hybrid tracking system

In this section, we have evaluated the tracking efficiency of our proposed hybrid method. The data sequences captured are of variable length ranging from 100 to 150 frames. **Figure** 7 shows results of tracking in different data sequences, approximately 10–12 frames of each data sequence are shown here to highlight the tracking efficiency of module II. Each frame is illustrated by its frame number, a yellow box enclosing the hand region and a yellow dot inside the yellow box represent the instant position of the centroid of the hand region. **Figure** 7(a) shows the tracking of P-I posture in a

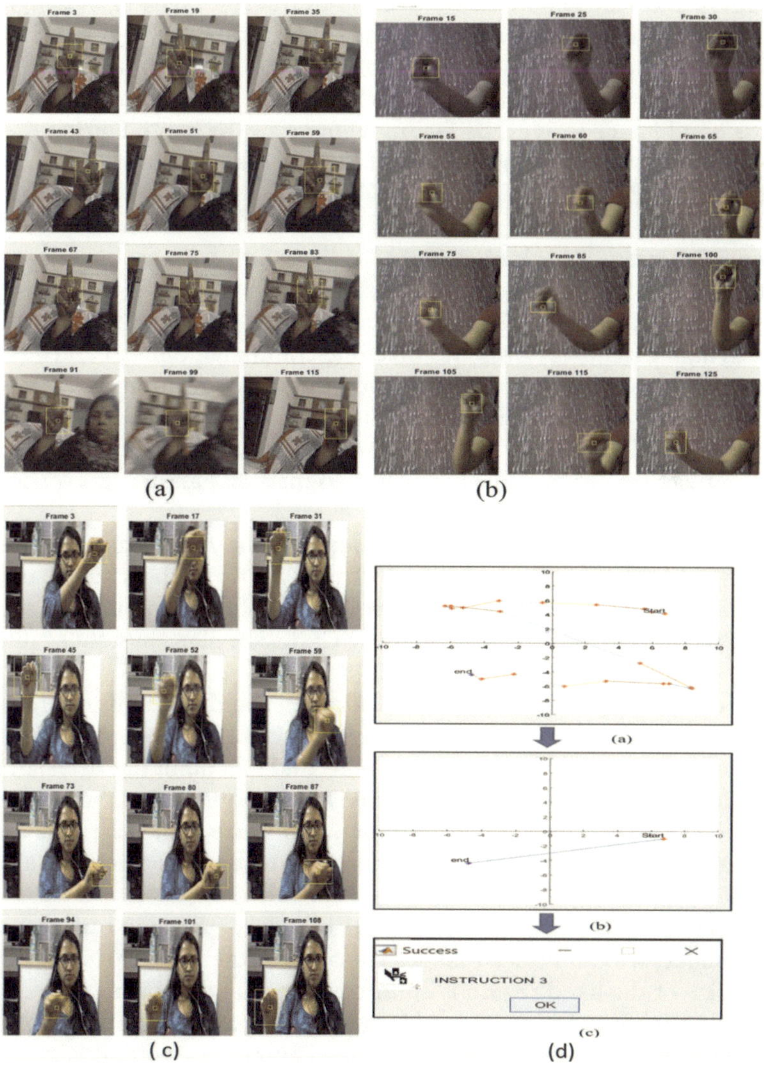

Figure 7.
Tracking outcomes of different data sequence are shown in (a), (b), (c), and (d) shows the cognitive recognition of hand movement in (c).

cluttered background. This data sequence is captured in a background that has many similar colored objects as that of the hand. Our proposed system discriminates and localizes the hand region efficiently due to the robust deep feature learning capability of our hybrid tracking system. It is also noticeable that the hand is properly identified even when the hand region was blurred due to sudden erratic movement by the subject as shown in frame 99 of the data sequence.

Figure 7(**b**) displays the tracking results of the P-III hand posture in improper illumination conditions. It can be noticed that in **Figure** 7(**a**) and (**b**), the FoS are frame 3 and frame 15, respectively. This data sequence is mainly affected by the color reflection of the background wall, and thus, it is visible that the edges of the P-III posture are nearly mixed with the background in some frames.

Figure 7(**c**) demonstrates the tracking results of a data sequence [32] in which a teenage girl is moving her hand (posture P-III) in front of her face. It is noticeable that the hand region and face region nearly overlap in frame 17. The fast change in the hand position in the frames indicates that the subject is moving her hand in a speedy manner. The change in the distance between the two positions of the hand frame 45 to frame 59 along with the change from a clear image of the hand region to the blurred image of the hand image proves the fast movement of the hand. During the movement, the subject is also changing the orientation of the hand posture as can be seen from the frames 59, 73, 80, 87.

5.4 Efficiency of hybrid tracking system

Cognitive efficiency means the development of the semantic between the trajectory of the dynamic hand gestures and machine command. Since, hand gestures do not follow a fixed line of movement to convey the same meaning. Therefore, syntax formation to match train data and test data is a challenge. Hence, the main limitation in DHGR is the development of a process that can convert the trajectory of hand movement to machine command. Our proposed method handles this difficult challenge in a schematic manner.

In our proposed technique, we have developed eight vision-based commands "INSTRUCTION 1–8" (abbreviated as INT-1 to INT-8). For the vision-based instruction, we have drafted a process to convert trajectory of the hand movement obtained in module-II to a machine command by using Cartesian plane system as illustrated in **Figure 8**.

Figure 7(d) illustrates the process in developing cognitive ability to recognize hand movement by the machine. This process consists of three steps: (i) trajectory plot of the hand movement, (ii) position of start and end point in Cartesian plane, and (iii) conversion to machine command. **Figure** 7(d) demonstrates the results of the cognitive recognition of a data sequence shown in **Figure** 7(**c**) [32]; here an adult girl moves her hand from right to left and the machine recognizes this movement as command 7.

Figure 9(**a**) shows tracking results of P-III posture performed by a teenage boy. In this data sequence, we can notice that scale change of the hand region is very prominent (as the size of the hand region is continuously changing from frame to frame). The posture area is big in frame 37, and it gradually decreases till frame 147. This indicates the distance between the subject's hand and the camera, it is minimum in frame 37 and maximum at frame 147. **Figure** 9(**b**) displays the result of cognitive recognition of the trajectory in the three steps in trajectory to command interpretation of left initiated data sequences The movement starts from the bottom left, moves in a

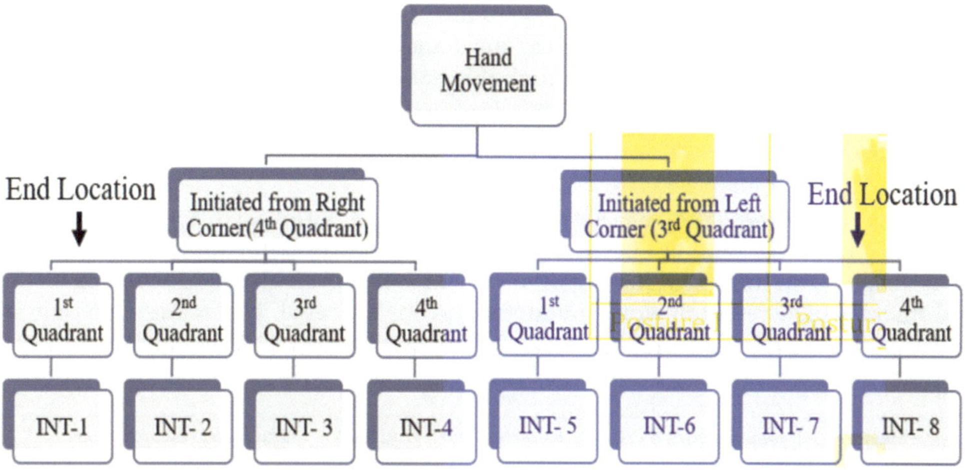

Figure 8.
Conversion of trajectory of hand movement to machine command.

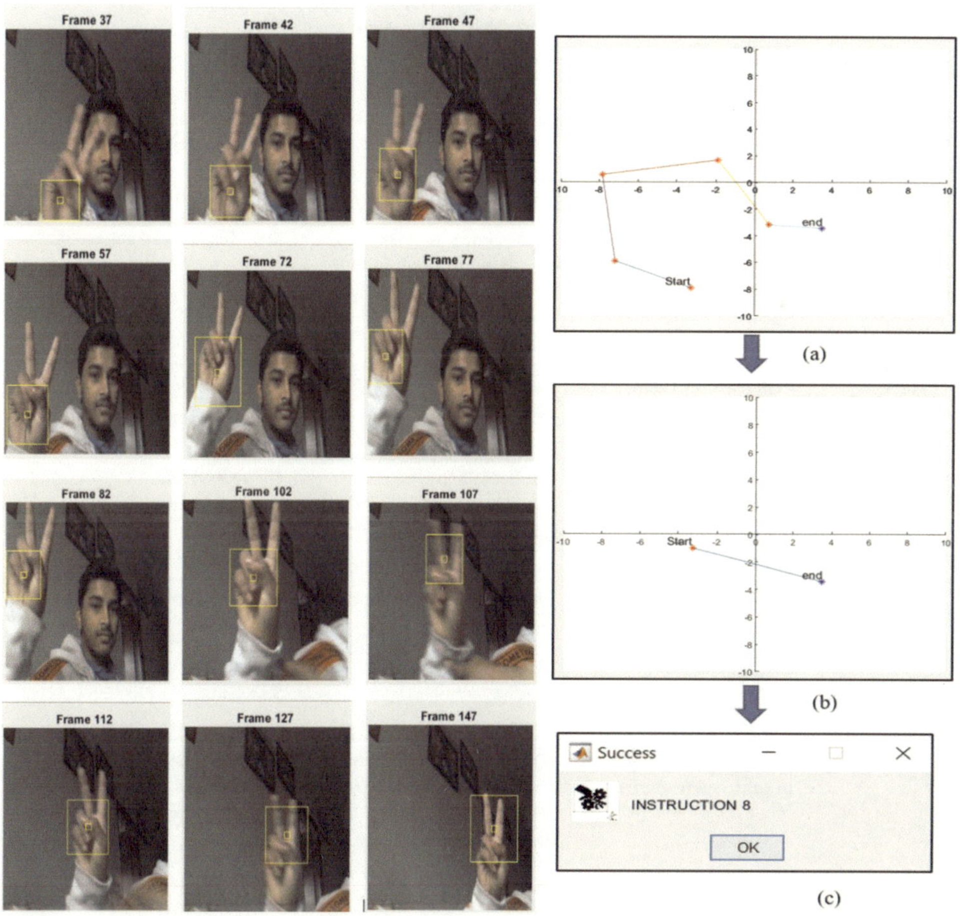

Figure 9.
Tracking results of P-III posture performed by a teenage boy. (b) Cognitive recognition of hand movement.

zigzag manner, and finally reaches close to the initial starting place. The PoS and end location of this sequence both are in the third and fourth quadrant respectively; thus, "INSTRUCTION 8" is generated through this hand movement.

5.5 Comparison with contemporary techniques

In this section, we compare our process and results with two different approaches used recently in the field of DHGR. In the first approach [32] technique utilizes true RGB images. This approach mainly involves hand-crafted features for hand detection and tracking. The research work conducted by Singha J. et al. [32] focused on only fist posture tracking in a fixed background, they have achieved 92.23% efficiency when no skin color object is present in the surroundings. One of the prominent limitations in

Parameters	Research work-I (2018) [36]	Research work-II (2020) [29]	Proposed research work
Camera/ Image type	Simple webcam/RGB	Microsoft Kinect Sensor version 2/depth,	Webcam/RGB
Preprocessing	Face segmentation using ViolaJones and the background subtraction using skin filtering	Noise Removal using median filtering and morphological processing. Conversion to binary image	Not Required
Initial stage-Hand detection	Three frames differencing on colored and grayscale images.	Hand Contour is extracted using Moore -Neighbor algorithm. Fingertip extraction using K-cosine algorithm.	Designed Faster-RCNN constructed on ResNet101. Used region-based network (RPN) for defining hand region.
Feature Extraction	Eigen features of the detected hand region. Remove unwanted features using compact criteria	Position of Fingertip calculated through inbuilt software of the camera.	SIFT feature extraction of AHT
Tracking methods	KLT features followed 44 features matching by compact criteria	For each frame, a 3D CNN is allotted.	Combination of Faster RCNN with SIFT algorithm.
Classification	Results of ANN, SVM, kNN classifiers are fused to get the final classified value	Ensemble learning to generate a final probability for classification	Using ANN with Cartesian quadrant system.
Background to conduct experiments	Fixed laboratory environment without any skin color object	Three fixed backgrounds	Any real-time background.
Accuracy of Methodology	92.23%	92.60%	95.83%
Limitations	KLT features get reduced in subsequent frames.	(i) Preprocessing is required (ii) For each frame separate 3D CNN is required this makes the system slow. (iii) fixed gesture length of 20 frames.	Initially trained for five gestures and can be extended for many more postures

Table 3.
Comparative analysis of two recent methods with the proposed methodology based on different parameters.

their approach is that they have applied sequence of algorithms for precise detection of hand region. This method is complex and unsuitable for real-time implementation of DHGR.

In the approach proposed by Tran DS et al. [29] for fingertip tracking, depth coordinates of fingertip provided by the inbuilt software of the advanced sensor-based camera are directly used. According to the researchers, RGB camera images are largely affected by illumination variation, and thus, to avoid background and illumination complexities in DHGR, they utilized RGB-D data sequences captured through the Microsoft Kinect V2 camera. It is a skeletal tracker camera that provides the position of 25 joints of the human skeleton including fingertips. This method is designed for tracking only seven hand movements comprised of 30–45 frames in three fixed backgrounds; besides, subjects are also trained to perform correct hand movement. In this research work, each frame is allotted an individual 3DCNN for classification. Thus, the experiments can perform fingertip tracking only for short gesture length. The training time of the 3D CNN is 1 hr. 35 minute with a six-core processor of 16GB RAM, which indicates the complex architecture of the technique. The accuracy of the trained 3D CNN model is 92.6% on validation data. **Table 3** illustrates and compares different technical aspect of the above two mentioned approaches with our proposed method:

6. Conclusion

This research work presents solutions to many crucial and unresolved challenges in vision-based tracking of hand movement captured using a simple camera. The methodology has the potential to provide a complete solution from hand detection to tracking and finally for cognitive recognition of trajectory to machine command for contactless Human-Machine interaction via dynamic hand gestures. Since the proposed design is implemented around a single RGB webcam, thus the system is economical and user-friendly. The accuracy achieved in the online and adaptive hand detection scheme with Faster R-CNN is 98.76%. The proposed hybrid tracking scheme exhibits high efficiency to adapt scale variation, illumination variation, and background conditions. It also exhibits high accuracy when camera is in motion during the movement. The overall accuracy achieved by our proposed system in complex conditions is 95.83%.

The comparative analysis demonstrates that our system gives users the freedom to select posture and to start the hand movement from any point in the image frame. Also, we do not impose any strict conditions in terms of geometrical shape of any posture. The hybrid framework and cognitive recognition features of our proposed method give a robust solution to classify any hand trajectory in a simple manner. This feature has not been discussed in any existing technique working with RGB images till date. The cumulative command interpretation efficiency of our system in real-time environment is 96.2%. The various results justify the "online" hand detection and "adaptive" tracking feature of the proposed technique. In the future, the method can be further extended to track multiple hand movements.

Acknowledgements

No funding is received

Author details

Richa Golash* and Yogendra Kumar Jain
Samrat Ashok Technological Institute, Vidisha, Madhya Pradesh, India

*Address all correspondence to: golash.richa@gmail.com

References

[1] Wachs JP, Kölsch M, Stern H, Edan Y. Vision-based hand-gesture applications. Communications of the ACM. 2011; **54**(2):60-71. DOI: 10.1145/1897816.1897838

[2] Golash R, Jain YK. Economical and user-friendly Design of Vision-Based Natural-User Interface via dynamic hand gestures. International Journal of Advanced Research in Engineering and Technology. 2020;**11**(6)

[3] Rautaray SS, Agrawal A. Vision based hand gesture recognition for human computer interaction: A survey. Artificial Intelligence Review. 2015; **43**(1):1-54. DOI: 10.1007/s10462-012-9356-9

[4] Mcintosh J. How it Works: BMW's Gesture Control. Available from: https://www.driving.ca/auto-news/news/how-it-works-bmw-gesture-control [Accessed: March 23, 2021]

[5] Gu C, Lien J. A two-tone radar sensor for concurrent detection of absolute distance and relative movement for gesture sensing. IEEE Sensors Letters. 2017;**1**(3):1-4. DOI: 10.1109/LSENS.2017.2696520

[6] Oudah M, Al-Naji A, Chahl J. Hand gesture recognition based on computer vision: A review of techniques. Journal of Imaging. 2020;**6**(8):73. DOI: 10.3390/jimaging6080073

[7] Li Y, Huang J, Tian F, Wang HA, Dai GZ. Gesture interaction in virtual reality. Virtual Reality & Intelligent Hardware. 2019;**1**(1):84-112

[8] Chakraborty BK, Sarma D, Bhuyan MK, MacDorman KF. Review of constraints on vision-based gesture recognition for human–computer interaction. IET Computer Vision. 2018; **12**(1):3-15

[9] Yasen M, Jusoh S. A systematic review on hand gesture recognition techniques, challenges and applications. PeerJ Computer Science. 2019; **16**(5):e218

[10] Golash R, Jain YK. Trajectory-based cognitive recognition of dynamic hand gestures from webcam videos. International Journal of Engineering Research and Technology. 2020;**13**(6): 1432-1440

[11] Yang H, Shao L, Zheng F, Wang L, Song Z. Recent advances and trends in visual tracking: A review. Neurocomputing. 2011;**74**(18):3823-3831

[12] Li X, Hu W, Shen C, Zhang Z, Dick A, Hengel AV. A survey of appearance models in visual object tracking. ACM Transactions on Intelligent Systems and Technology (TIST). 2013;**4**(4):1-48

[13] Bandini A, Zariffa J. Analysis of the hands in egocentric vision: A survey. IEEE Transactions on Pattern Analysis and Machine Intelligence. 2020

[14] Golash R, Jain YK. Robust tracking of moving hand in coloured video acquired through simple camera. International Journal of Computer Applications in Technology. 2021;**65**(3): 261-269

[15] Bandara HM, Priyanayana KS, Jayasekara AG, Chandima DP, Gopura RA. An intelligent gesture classification model for domestic wheelchair navigation with gesture variance compensation. Applied Bionics and Biomechanics. 2020;**30**:2020

[16] Wang J, Payandeh S. Hand motion and posture recognition in a network of calibrated cameras. Advances in Multimedia. 2017;**2017**:25. Article ID 216207. DOI: 10.1155/2017/2162078

[17] Poon G, Kwan KC, Pang WM. Real-time multi-view bimanual gesture recognition. In: 2018 IEEE 3rd International Conference on Signal and Image Processing (ICSIP). IEEE; 2018. pp. 19-23

[18] Cruz Bautista AG, González-Barbosa JJ, Hurtado-Ramos JB, Ornelas-Rodriguez FJ, González-Barbosa EA. Hand features extractor using hand contour–a case study. Automatika. 2020; **61**(1):99-108

[19] Marin G, Dominio F, Zanuttigh P. Hand gesture recognition with jointly calibrated leap motion and depth sensor. Multimedia Tools and Applications. 2016;**75**(22):14991-15015

[20] Kainz O, Jakab F. Approach to hand tracking and gesture recognition based on depth-sensing cameras and EMG monitoring. Acta Informatica Pragensia. 2014;**3**(1):104-112

[21] Aristidou A. Hand tracking with physiological constraints. The Visual Computer. 2018;**34**(2):213-228

[22] Abraham L, Urru A, Normani N, Wilk MP, Walsh M, O'Flynn B. Hand tracking and gesture recognition using lensless smart sensors. Sensors. 2018; **18**(9):2834

[23] Huang H, Chong Y, Nie C, Pan S. Hand gesture recognition with skin detection and deep learning method. Journal of Physics: Conference Series. 2019;**1213**(2):022001

[24] Yao MH, Gu QL, Wang XB, He WX, Shen Q. A novel hand gesture tracking algorithm fusing Camshift and particle filter. In: 2015 International Conference on Artificial Intelligence and Industrial Engineering. Atlantis: Atlantis Press; 2015

[25] Khaled H, Sayed SG, Saad ES, Ali H. Hand gesture recognition using modified 1$ and background subtraction algorithms. Mathematical Problems in Engineering. 2015;**20**:2015

[26] Liu P, Li X, Cui H, Li S, Yuan Y. Hand gesture recognition based on single-shot multibox detector deep learning. Mobile Information Systems. 2019;**30**:2019

[27] Bao P, Maqueda AI, del-Blanco CR, García N. Tiny hand gesture recognition without localization via a deep convolutional network. IEEE Transactions on Consumer Electronics. 2017;**63**(3):251-257

[28] Shin J, Kim H, Kim D, Paik J. Fast and robust object tracking using tracking failure detection in kernelized correlation filter. Applied Sciences. 2020;**10**(2):713

[29] Tran DS, Ho NH, Yang HJ, Baek ET, Kim SH, Lee G. Real-time hand gesture spotting and recognition using RGB-D camera and 3D convolutional neural network. Applied Sciences. 2020;**10**(2):722

[30] Zhao D, Liu Y, Li G. Skeleton-based dynamic hand gesture recognition using 3d depth data. Electronic Imaging. 2018; **2018**(18):461-461

[31] Plouffe G, Cretu AM. Static and dynamic hand gesture recognition in depth data using dynamic time warping. IEEE Transactions on Instrumentation and Measurement. 2015;**65**(2):305-316

[32] Singha J, Roy A, Laskar RH. Dynamic hand gesture recognition using

vision-based approach for human–computer interaction. Neural Computing and Applications. 2018;**29**(4):1129-1141

[33] Barros P, Maciel-Junior NT, Fernandes BJ, Bezerra BL, Fernandes SM. A dynamic gesture recognition and prediction system using the convexity approach. Computer Vision and Image Understanding. 2017; **1**(155):139-149

[34] Ren S, He K, Girshick R, Sun J. Faster R-CNN: Towards real-time object detection with region proposal networks. IEEE Transactions on Pattern Analysis and Machine Intelligence. 2016;**39**(6): 1137-1149

[35] He K, Zhang X, Ren S, Sun J. Deep residual learning for image recognition. In: Proceedings of the IEEE Conference on Computer Vision and Pattern Recognition. IEEE; 2016. pp. 770-778

[36] Lowe DG. Distinctive image features from scale-invariant keypoints. International Journal of Computer vision. 2004;**60**(2):91-110

[37] Lindeberg T. Scale invariant feature transform. Scholarpedia. 2012;7(5): 10491

[38] Rojas R. Fast learning algorithms, in neural networks. Springer. 1996;**1996**: 183-225

Section 2

Information Extraction

Chapter 5

Thresholding Image Techniques for Plant Segmentation

Miguel Ángel Castillo-Martínez,
Francisco Javier Gallegos-Funes, Blanca E. Carvajal-Gámez,
Guillermo Urriolagoitia-Sosa and Alberto J. Rosales-Silva

Abstract

There are challenges in the image-based research to obtain information from the objects in the scene. Moreover, an image is a set of data points that can be processed as an object in similarity way. In addition, the research fields can be merged to generate a method for information extraction and pixel classification. A complete method is proposed to extract information from the data and generate a classification model capable to isolate those pixels that are plant from others are not. Some quantitative and qualitative results are shown to compare methods to extract information and create the best model. Classical and threshold-based state-of-art methods are grouped in the present work for reference and application in image segmentation, obtaining acceptable results in the plant isolation.

Keywords: similarity, classification, threshold, image processing, segmentation

1. Introduction

There are three fields for the image-based research: Image processing, Computer Vision and Computer Graphics [1]. In a graphical way, this is shown in the **Figure 1**.

Image processing takes an image as input and realize a set of operations to create a new image that improves the interest feature visualization. It is not limited to; the image processing can isolate those features have not meaningful information to be removed from the scene. For Cancer Aided Diagnostic in dermoscopy, shown in **Figure 2**, the lesion must be bounded but there are meaningless elements as hair and air bubbles. Here, there is a need to improve the lesion visualization removing all those elements. An approach for the processing chain begins with color space transformation, continues with hair detection and finishes with image inpainting [2].

Computer Vision starts with an image and provides features of the object in the scene. These features allow a quantitative description for object interpretation. The **Figure 3** takes the region of interest (ROI), hand for this case, and the 7 Hu's moments are calculated to describe the hand sign as a feature vector, simplifying the classification process [3].

Computer graphics generates a visual representation of mathematical model behavior. The models can be visualized from lines to a video that shows time evolve

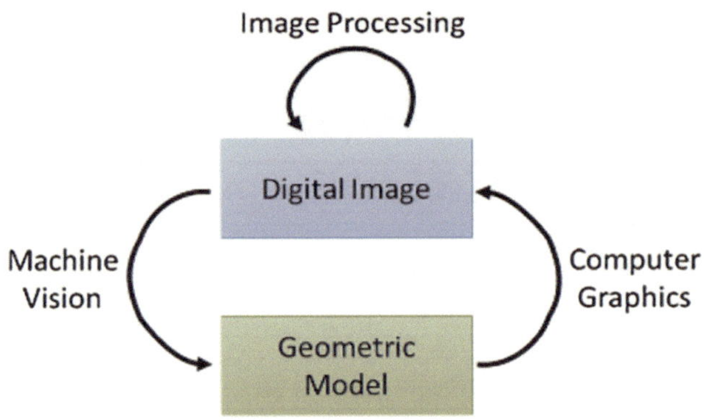

Figure 1.
Fields in image-based research. Source: [1].

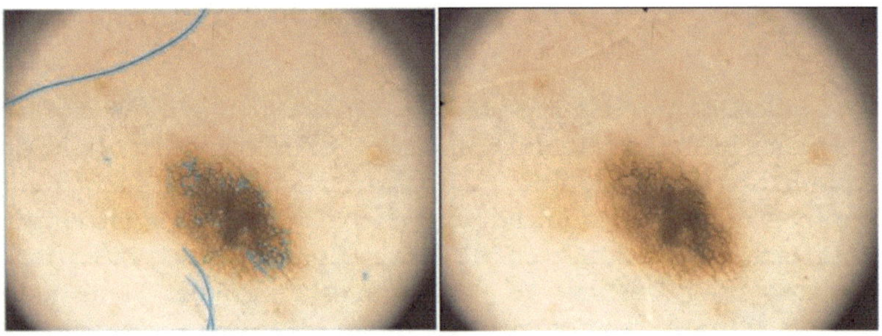

Figure 2.
Hair Inpainting in dermoscopic images.

Hu Moments	A
ϕ_1	0.69
ϕ_2	2.03
ϕ_3	2.83
ϕ_4	3.46
ϕ_5	6.68
ϕ_6	4.56
ϕ_7	7.44

Figure 3.
Static sign recognition. Source: [3].

behaviors. **Figure 4** shows a finite differences description to represent a coil with ferromagnetic core.

The applications are not exclusive from each field of study, these fields could be merged to generate a hybrid system that improves the description and composition to solve a specific problem. To hair removal in the **Figure 2**, a new image with the hair, described by a statistical operator and a thresholding rule, is generated. The inpainting takes the original and generated images to process only the pixels that do not describe the lesion, this process minimizes the classification error. The solution employees three research fields in image-based systems.

Color indexes. For the primary color image encoding the RGB color space is used. This allows the storage and image representation in Red, Green and Blue colors [5–7].

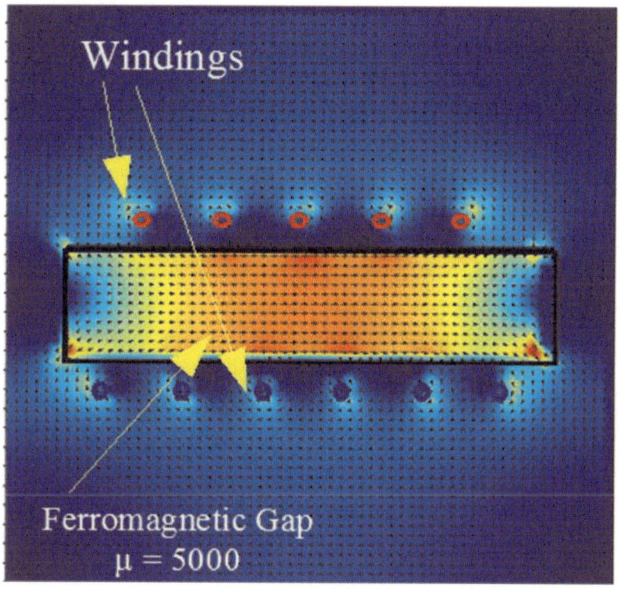

Figure 4.
Numerical analysis for non-destructive electromagnetic test. Source: [4].

Transform the color space brings other image description to simplify the processing or improve visualization.

In matrix representation, the color transformation has the following form,

$$I = T_S \bullet p + k \tag{1}$$

where I is the color space with N components, T_S is the $N \times 3$ transformation matrix, p is the RGB vector representation of the pixel, and k is a constant vector. This representation allows a generalized transformation for N Channels with N transformation equations.

Suppose the RGB to YUV color space transformation [5],

$$\begin{aligned} Y &= 0.299R + 0.587G + 0.114B \\ U &= 0.492(B - Y) \\ V &= 0.877(R - Y) \end{aligned} \tag{2}$$

Substituting and expanding Eq. (2),

$$\begin{aligned} Y &= 0.299R + 0.587G + 0.114B \\ U &= -0.147R - 0.289G + 0.436B \\ V &= 0.615R - 0.515G - 0.100B \end{aligned} \tag{3}$$

For this case, k is a zeros vector. This allows represent RGB to YUV as matrices with Eq. (4)

$$\begin{pmatrix} Y \\ U \\ V \end{pmatrix} = \begin{pmatrix} 0.299 & 0.587 & 0.114 \\ -0.147 & -0.289 & 0.436 \\ 0.615 & -0.515 & -0.100 \end{pmatrix} \begin{pmatrix} R \\ G \\ B \end{pmatrix} \tag{4}$$

Because color transformations do not consider neighborhood pixels, these are known as point-to-point operations. The color spaces are described for break up color and lighting. Moreover, there are other methods to transform color spaces and simplify feature extraction in images, introducing the color index concept.

A color index can take the color space information and generates a new channel, this improves the feature visualization according to the requirements. For plant images, Yuan et al. estimates the Nitrogen content in rice from plant in the digital image [8]. Authors get a measure called GMR, see a representation in **Figure 5**, subtracting the red channel from green channel responses respectively. After, they apply a fixed threshold for the plant segmentation.

The Color Index Vegetation Extraction (CIVE) is used to break up plants and soil. This allows a grow evaluation in the crops. Furthermore, the CIVE shows good response in outdoor environments [9, 10]. If the color is processed in the GMR and CIVE, the color transformation is defined as,

$$\begin{pmatrix} GMR \\ CIVE \end{pmatrix} = \begin{pmatrix} -1 & 1 & 0 \\ 0.441 & -0.811 & 0.385 \end{pmatrix} \begin{pmatrix} R \\ G \\ B \end{pmatrix} + \begin{pmatrix} 0 \\ 18.78745 \end{pmatrix} \tag{5}$$

Similarity Measure. Minkowski distance is a generalized way to similarity measure [11–14] defined as,

$$d_n(X, Y) = \left(\sum_{i=1}^{n} \left(|x_i - y_i| \right)^n \right)^{1/n} \tag{6}$$

where $X = \{x_1, x_2, \cdots, x_n\}, Y = \{y_1, y_2, \cdots, y_n\} \in \mathcal{R}^n$ are data points which the algorithm seeks minimum distance. If $n = 2$ then Euclidean distance is measured. Substituting in Eq. (6) the following expression is obtained,

$$d_2(X, Y) = \sqrt{(X - Y)^{\intercal}(X - Y)} \tag{7}$$

Other way to measure similarity is by Mahalanobis distance [11, 15], this is calculated by Eq. (8),

$$d_M(x, y) = \sqrt{(X - Y)^{\intercal}\Sigma^{-1}(X - Y)} \tag{8}$$

where Σ is the covariance matrix.

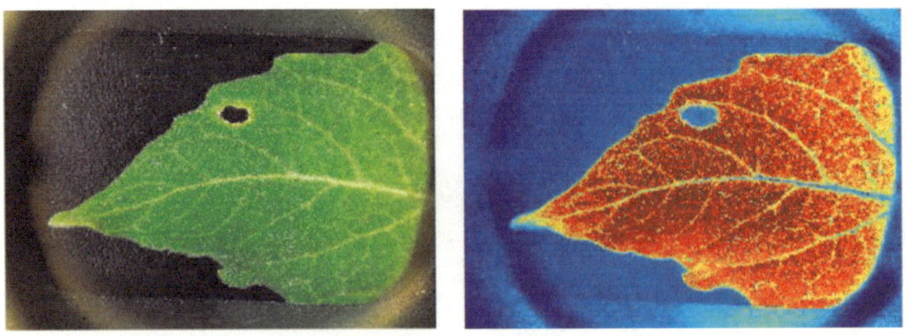

Figure 5.
Colored GMR response of a leaf in an image acquired with polarized light.

$$\Sigma = \begin{bmatrix} \sigma_{11}^2 & \sigma_{12}^2 & \cdots & \sigma_{1n}^2 \\ \sigma_{21}^2 & \sigma_{22}^2 & \cdots & \sigma_{2n}^2 \\ \vdots & \vdots & \ddots & \vdots \\ \sigma_{n1} & \sigma_{n2}^2 & \cdots & \sigma_{nn}^2 \end{bmatrix} \tag{9}$$

If $\Sigma = I$ then Eq. (8) brings the Euclidean distance. Thus, a Weighted Euclidean distance can be calculated as

$$d_\Omega(x,y) = \sqrt{(X-Y)^\mathsf{T}\Omega^{-1}(X-Y)} \tag{10}$$

where Ω is a $n \times n$ weight matrix. If each component is independent form others, then is possible define a weight matrix $W = \Sigma\mathrm{I}$.

$$W = \begin{bmatrix} \sigma_{11}^2 & 0 & \cdots & 0 \\ 0 & \sigma_{22}^2 & \cdots & 0 \\ \vdots & \vdots & \ddots & \vdots \\ 0 & 0 & \cdots & \sigma_{nn}^2 \end{bmatrix} \tag{11}$$

With the analysis above, there are 3 cases for the weight matrix:

1. $\Omega = \Sigma$: Mahalanobis distance is calculated

2. $\Omega = I$: Euclidean distance is calculated

3. $\Omega = W$: Weighted Euclidean distance is calculated.

Find features that describe the object of interest is required to calculate the similarity for each pixel in the image. The distance measure brings the similarity between the pixel and plants, background, fruit and more.

Thresholding. Thresholding is based on a simple rule to cluster data in groups k_0 and k_1 from a threshold T. The groups have not a meaning in their clustering. Moreover, it is possible use this rule as supervised method (classification) [16–19]. The rule R to identify a data point is described below.

$$R = \begin{cases} k_0, & \varepsilon < T \\ k_1, & otherwise \end{cases} \tag{12}$$

where ε is a measure according to analysis.

For digital images, an example for pixel identification is shown in **Figure 6**. From this figure, T is selected with the mean of the data and the pixels are assigned to a group according to it.

There are two challenges using this rule: which parameters define ε and what conditions bring T. According to the analysis, ε could be an independent component. This means that the rule can use only one-color feature. Furthermore, there is a possibility to make a feature composition, improving results. In other perspective, T can be defined by a measure that describes data features for ε and allow an acceptable division between groups.

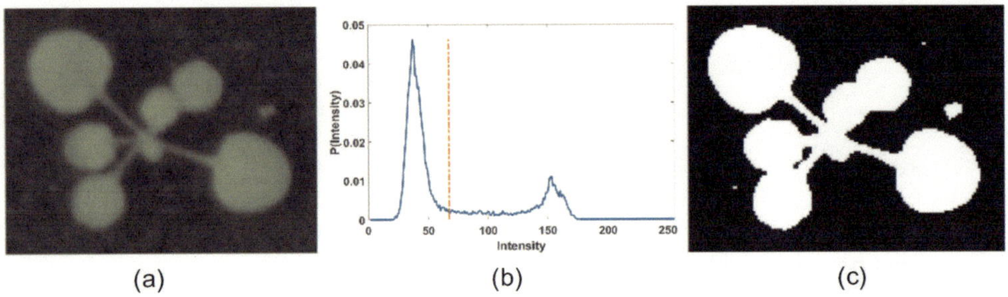

Figure 6.
Detected ROI by mean thresholding. a) Original image (Green Channel). b) Histogram (blue) and mean value (red). c) Grouped pixels.

In state-of-art methods for threshold calculation, the Otsu method is reached [20]. This method brings thresholds that have the better separability over the data. Some indexes and separability obtained are shown in **Figure 7**.

In order to measure the quality in the data isolation, the information gain is used. The information gain, good for decision tree generation, is a data homogeneity indicator [21–23]. This allows a class probability measure for a dataset distribution, indicating that all classes have the same probability when entropy is equal to one.

Information gain is calculated by Eq. (13)

$$G(X|T_n) = H(X) - \sum_{v \in T} \frac{|X|T_v|}{|X|} \bullet H(X|T_v) \tag{13}$$

where v is each possible value in the random variable for the analyzed color index T_n, $X|T_v$ is the sub generated dataset for T_n,$|\bullet|$ is the dataset cardinality and $H(X)$ is dataset entropy.

For threshold case,

$$G(X|T_n) = H(X) - \left(\frac{|X|T_<|}{|X|} \bullet H(X|T_<) + \frac{|X|T_\geq|}{|X|} \bullet H(X|T_\geq) \right) \tag{14}$$

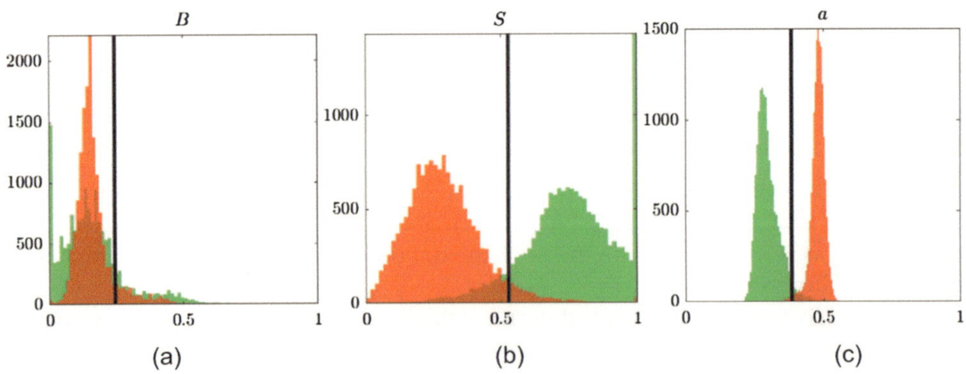

Figure 7.
CVPPP color index pixel generated dataset break up. a) B Channel from RGB space: There is no way to separate data; b) S Channel from HSV space: There is a medium separability; c) a channel from lab space: There is an acceptable break up. Green: Plant, red: Background, histogram: 64 levels.

Entropy is a random variable uncertainty measure. This is the most utilized information measure in this kind of processes [24–26]. In a normalized way, the Shannon entropy is calculated as follows

$$H(X) = -\sum_{k=1}^{C} P_k \bullet \log_C(P_k) \tag{15}$$

where C is the available classes in the dataset and P_k is the corresponding probability for each class.

Considering M pixels and each class has M/C elements, Eq. (15) changes to,

$$H(X) = -\sum_{k=1}^{C} \frac{1}{C} \bullet \log_C\left(\frac{1}{C}\right) \tag{16}$$

For the binary case, C is replaced by 2, then

$$H(X) = -\sum_{k=1}^{2} \frac{1}{2} \bullet \log_2\left(\frac{1}{2}\right) = 1 \tag{17}$$

If $H(X) = 1, C > 1$ then all classes have the same probability. This means that the dataset distribution for feature selection is uniform.

If the conditional entropy $H(X|T_<) = H(X|T_\geq) = 0$ then each sub dataset has elements that belongs to only one class k, this is a total separability indicator.

2. Methodology

A proposed method consists in two main sections: Feature selection and classification. Feature selection takes the better features that break up those pixels in foreground and background. Classification utilizes a similarity measure to compare pixels and plant description to assign a specific class. Both processes are described below.

To select features that belong to plants in the best way, the method requires a similarity measure. This measure allows to compare and identify which can provide an acceptable separation between data. Considering a dataset with the same quantity samples of each class, the dataset has the form described in **Figure 8**.

The feature i_n is considered as continuous random variable. In order to simplify and use simple tools in the classification model generation, the random variable is discretized to a binary variable. Thus, the dataset is presented in **Figure 9**.

For data classification, the minimum distance is wanted between the pixel and the object of interest. This analysis requires those features that describes the object to be compared with those features that describes the pixel. For plant image processing, some features are the color indexes as NCIVE, MNGRDI, GMR, etc. The feature selection is dependent to those maximizes data separation.

When the features are defined, independent statistical features are calculated. This information allows an orientation correction and distance threshold magnitude computation. According to this analysis, all components are considered independent from others, thus, the distance measure takes the third scenario. This weights those features that have more variability and brings an eccentricity adjust. An orientation correction should be applied to data because Euclidean distance is rotation variant.

i_1	i_2	$\cdots$	i_N	$Class$
x_1^1	x_2^1	$\cdots$	x_N^1	F
x_1^2	x_2^2	$\cdots$	x_N^2	F
$\vdots$	$\vdots$	$\ddots$	$\vdots$	$\vdots$
x_1^M	x_2^M	$\cdots$	x_N^M	B

Figure 8.
Pixel dataset representation. Where i es the feature vector that describes the object under study. The dataset contains N features, and Class describes what is the meaning of the feature vector, in this case, foreground or background.

T_1	T_2	$\cdots$	T_N	$Class$
$<$	$<$	$\cdots$	$\geq$	F
$<$	$\geq$	$\cdots$	$<$	F
$\vdots$	$\vdots$	$\ddots$	$\vdots$	$\vdots$
$\geq$	$<$	$\cdots$	$<$	B

Figure 9.
Discretized variable dataset representation. Where T_n is a threshold overcome indicator.

For threshold calculation the standard deviations $\sigma_{11}, \sigma_{22}, \cdots \sigma_{nn} \in \mathcal{R}^n$ are considered as a component vector. The magnitude is calculated with the Weighted Euclidean Distance and is assigned to the threshold. The standard deviation is considered because variance cannot be expressed in the same plane. This thought is expressed as follows,

$$Th = d_\Omega(0, \sigma)|\Omega = W \tag{18}$$

Before calculating distances, an orientation correction is needed with the angle described by the data. Finally, the thresholding defines if a pixel is plant or not with the following rule

$$Plant_{(r,c)} = \begin{cases} True & d_W\left(p_{(r,c)}, R\right) < Th \\ False & otherwise \end{cases} \tag{19}$$

where $Plant_{(r,c)}$ is the classification as plant for pixel $p_{(r,c)}$ and R is the feature vector defined by data in plant class. As result, the classification model with this method is shown in **Figure 10**.

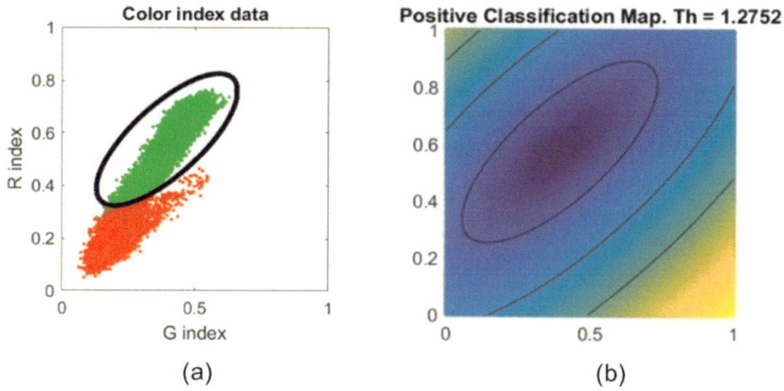

Figure 10.
Pixel classification map. a) Data classification with decision boundary in Th; b) normalized classification map with decision boundary in [Th, 2Th, 3Th]. Green: Plant, red: Background, black: Boundary decision. Th: Computed threshold.

3. Results

In the following processes the MATLAB 2020b was used in a Laptop with Xeon E3-1505 M processor with32GB RAM. In addition, only the matrix operations were used for data processing and data image representation.

According to the study in [27], the optimal color index for data break up is the L and a channels in the Lab color space. Moreover, some experiments are done according the dataset variability from the CVPPP dataset [28], which is the test dataset. In one hand, comparing both results is observable that a channel matches with [27] as best index to isolate the plant pixels. In the other hand, there are other indexes that provide an acceptable separability, NCIVE and MNGRDI, shown in **Tables 1** and **2**. In a comparative way, the Otsu method thresholding has a separability less than the supervised entropy.

i	Threshold	Information Gain
a	0.3843	0.8421
NCIVE	0.4196	0.8036
MNGRDI	0.5294	0.7537
G	0.3882	0.6913
b	0.6705	0.6729
L	0.3843	0.6601
V	0.4078	0.6387
S	0.5294	0.6014
R	0.3137	0.4325
H	0.5098	0.1597
B	0.247	0.0074

Table 1.
Indexes gain information. Otsu method.

i	Threshold	Information Gain
a	0.3968	0.8471
NCIVE	0.4444	0.8428
MNGRDI	0.5238	0.7467
G	0.3174	0.7265
V	0.4078	0.6894
b	0.6507	0.6786
L	0.3333	0.6601
S	0.492	0.6055
R	0.2857	0.4489
H	0.1904	0.3629
B	0.063	0.0748

Table 2.
Indexes gain information. Supervised entropy.

Other observation is the similarity in the order of separability quality. Moreover, the supervised entropy knows the data meaning, consequently, the quality measure is better in most of the cases.

In **Figure 11** some examples of the index threshold data separability are shown. From here, is observed twice best cases, with an acceptable separability, and twice worst cases, with no apparent separability way.

Finally, Classification map for plant description pixels based on a-NCIVE indexes is shown in the **Figure 12**.

According to this model, the pixels can be classified as plant or not. Some visual results are shown in the **Figure 13**.

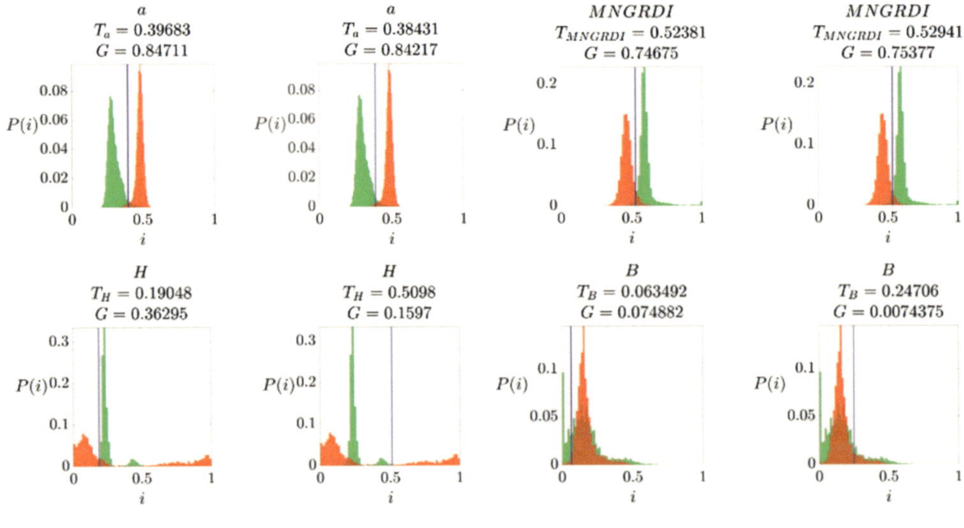

Figure 11.
Thresholds and information gains for some color indexes. Left: Supervised entropy method, right: Otsu method.

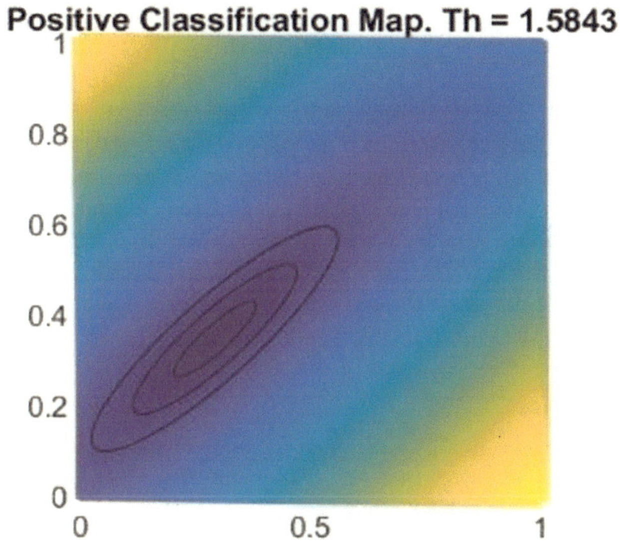

Figure 12.
Classification map for plant pixel segmentation. Th: Computed threshold.

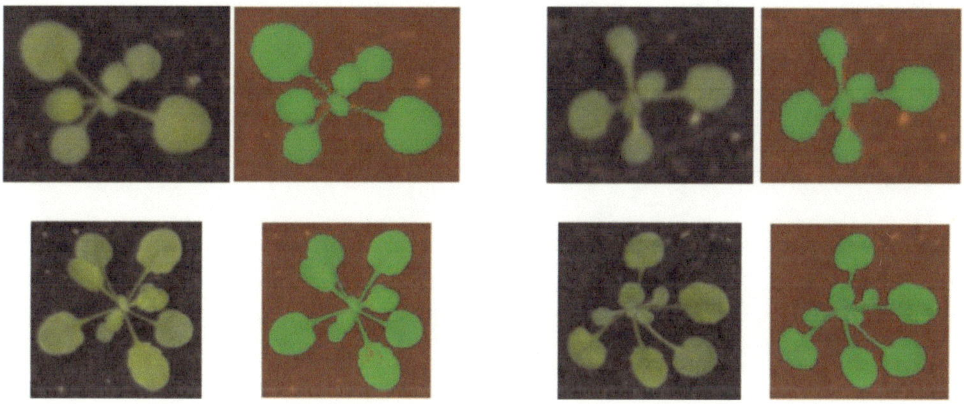

Figure 13.
Visual results in plant image segmentation for CVPPP dataset.

4. Conclusions

The thresholding methods are effective when the problem data is defined and the superposition between groups is minimum. Furthermore, they are simple methods that provide acceptable results in the segmentation problem. The combination of methods is possible to rise the quality in the models.

The entropy, in a supervised way, can improve the data separability. Because Otsu methods minimizes the variance between groups, the quality in the results using supervised entropy is improved in consequence of consider data meaning. The supervised methods know the expected response, breaking up data in corresponding classes. Otsu only allow clustering data that have not a well-defined meaning yet.

The distance measures refer to reach a similarity in the compared data. It only needs a reference description of the studied object. In this case, all pixels are defined as plant, in a statistical way, a reference with the new pixels going to compared and classified. Thus, the segmentation is possible on those pixels that belong to plant and discard those are not.

From **Tables 1** and **2**, the best indexes have the same order in both results. Furthermore, the calculated threshold improves the data separability for the supervised entropy case. This allows development of classification maps like in **Figure 12** to consider those indexes that achieve the best pixels break up. **Figure 11** shown some separation scenarios where pixels that are plant (Green distribution) and are not (Red distribution). Finally, the classification map, shown in **Figure 12**, illustrates the best classifier obtained with the method to select pixels that belong to plant class.

Acknowledgements

To M. Minervi et al. for providing their dataset. This work is supported by Instituto Politecnico Nacional de Mexico (IPN) (Grant ID: 20201681) and Consejo Nacional de Ciencia y Tecnologia (CONACyT), project 240820.

Conflict of interest

The authors declare no conflict of interest.

Author details

Miguel Ángel Castillo-Martínez*, Francisco Javier Gallegos-Funes,
Blanca E. Carvajal-Gámez, Guillermo Urriolagoitia-Sosa and Alberto J. Rosales-Silva
Escuela Superior de Ingeniería Mecánica y Eléctrica, Instituto Politécnico Nacional, Col. Lindavista, Ciudad de Mexico, Mexico

*Address all correspondence to: macastillom@ipn.mx

References

[1] Hunt K. Introduction. In: The Art of Image Processing with Java. United States: A K Peters/CRC Press; 2010. pp. 1–12. Available from: https://doi.org/10.1201/9781439865590

[2] Castillo Martínez MA, Gallegos Funes FJ, Rosales Silva AJ, Ramos Arredondo RI. Preprocesamiento de imágenes dermatoscopicas para extracción de características. Research Computers Science. 2016;**114**(1):59-70 [Internet] Available from: http://rcs.cic.ipn.mx/2016_114/Preprocesamiento de imagenes dermatoscopicas para extraccion de caracteristicas.pdf

[3] Pérez LM, Rosales AJ, Gallegos FJ, Barba AV. LSM static signs recognition using image processing. In: 14th International Conference on Electrical Engineering, Computing Science and Automatic Control (CCE) [Internet]. Mexico City: IEEE; 2017. pp. 1-5. Available from: http://ieeexplore.ieee.org/document/8108885/

[4] Chávez-González AF, Aguila-Munoz J, Perez-Benitez JA, Espina-Hernandez JH. Finite differences software for the numeric analysis of a non-destructive electromagnetic testing system. In: 23rd International Conference on Electronics, Communications and Computing [Internet]. Cholula: IEEE; 2013. pp. 82-86. Available from: http://ieeexplore.ieee.org/document/6525764/

[5] Burger W, Burge MJ. Color images. In: Digital Image Processing. London: Springer; 2nd edition, 2016. pp. 291–328. Available from: http://link.springer.com/10.1007/978-1-4471-6684-9_12

[6] Sundararajan D. Color image processing. In: Digital Image Processing [Internet]. Singapore: Springer Singapore; 2017. pp. 407-438 Available from: http://link.springer.com/10.1007/978-981-10-6113-4_14

[7] Gonzalez RC, Woods RE. Color image processing. In: Digital Image Processing. England: Pearson; 2018. 4th edition, pp. 399-461

[8] Wang Y, Wang D, Zhang G, Wang J. Estimating nitrogen status of rice using the image segmentation of G-R thresholding method. Field Crops Research. 2013;**149**: 33-39. [Internet]. Available from:. DOI: 10.1016/j.fcr.2013.04.007

[9] Hamuda E, Glavin M, Jones E. A survey of image processing techniques for plant extraction and segmentation in the field. Computer and Electronics Agriculture. 2016;**125**:184-199. [Internet]. Available from:. DOI: 10.1016/j.compag.2016.04.024

[10] Kataoka T, Kaneko T, Okamoto H, Hata S. Crop growth estimation system using machine vision. Proceedings of 2003 IEEE/ASME International Conference on Advanced Intelligent Mechatronics (AIM 2003). 2003;**2** (Aim):1079-1083

[11] Flach P. Distance-based models. In: Machine Learning [Internet]. Cambridge: Cambridge University Press; 2012. pp. 231-261. Available from: https://www.cambridge.org/core/product/identifier/CBO9780511973000A014/type/book_part

[12] Aldridge M. Clustering: An overview. In: Lecture Notes in Data Mining [Internet]. London: World Scientific; 2006. pp. 99-107. Available from: http://www.worldscientific.com/doi/abs/10.1142/9789812773630_0009

[13] Zhou ZH. Clustering. In: Machine Learning [Internet]. Singapore: Springer

Singapore; 2021. pp. 211-240. Available from: https://link.springer.com/10.1007/978-981-15-1967-3_9

[14] Zhang D. Image ranking. In: Fundamentals of Image Data Mining. Cham: Springer; 2019. p. 271-287. Available from: http://link.springer.com/10.1007/978-3-030-17989-2_12

[15] Zhang Y, Li Z, Cai J, Wang J. Image segmentation based on FCM with mahalanobis distance. In: International Conference on Information Computing and Applications. Berlin: Springer; 2010. pp. 205-212. Available from: http://link.springer.com/10.1007/978-3-642-16167-4_27

[16] Flach P. Machine Learning [Internet]. 2nd ed. Cambridge: Cambridge University Press; 2012. Available from: http://ebooks.cambridge.org/ref/id/CBO9780511973000

[17] Sundararajan D. Digital Image Processing [Internet]. Singapore: Springer Singapore; 2017. Available from: https://link.springer.com/book/10.1007/978-981-10-6113-4

[18] Burger W, Burge MJ. Point Operations. In: Digital image processing [internet]. 2nd ed. London: Springer; 2016:57-88. Available from: https://link.springer.com/chapter/10.1007/978-1-4471-6684-9_4

[19] Gonzalez RC, Woods RE, Eddins SL. Image segmentation I. In: Digital Image Processing Using MATLAB. 3rd ed. United States of America: Gatesmark Publishing; 2020. pp. 633-721

[20] Otsu N. A threshold selection method from gray-level histograms. IEEE Transactions on Systems, Man, and Cybernetics. 1979;**9**(1):62-66 [Internet] Available from: http://ieeexplore.ieee.org/document/4310076/

[21] Rokach L, Maimon O. Splitting criteria. In: Data Mining with Decision Trees [Internet]. Singapore: World Scientific Publishing; 2014. pp. 61-68. Available from: http://www.worldscientific.com/doi/abs/10.1142/9789814590082_0005

[22] Masud MM, Khan L, Thuraisingham B. Email worm detection using data mining. In: Techniques and Applications for Advanced Information Privacy and Security [Internet]. Hershey, PA: IGI Global; 2009. pp. 20-34. Available from: http://services.igi-global.com/resolvedoi/resolve.aspx?doi=10.4018/978-1-60566-210-7.ch002

[23] Omitaomu OA. Decision trees. In: Lecture Notes in Data Mining [Internet]. London: World Scientific; 2006. pp. 39-51. Available from: http://www.worldscientific.com/doi/abs/10.1142/9789812773630_0004

[24] Addison PS. The Illustrated Wavelet Transform Handbook [Internet]. Biomedical Instrumentation and Technology. Boca Raton: CRC Press; 2017. p. 163. Available from: https://www.taylorfrancis.com/books/9781315372556

[25] Rubinstein RY, Kroese DP. The Cross-Entropy Method [Internet]. New York, NY: Springer New York; 2004 (Information Science and Statistics). Available from: http://link.springer.com/10.1007/978-1-4757-4321-0

[26] Ito S, Sagawa T. Information flow and entropy production on Bayesian networks. In: Mathematical Foundations and Applications of Graph Entropy [Internet]. Weinheim, Germany: Wiley-VCH Verlag GmbH & Co. KGaA; 2016. pp. 63-99 Available from: http://doi.wiley.com/10.1002/9783527693245.ch3

[27] Hernández-Hernández JL, García-Mateos G, González-Esquiva JM,

Escarabajal-Henarejos D, Ruiz-Canales A, Molina-Martínez JM. Optimal color space selection method for plant/soil segmentation in agriculture. Computers and Electronics in Agriculture. 2016;**122**: 124-132

[28] Minervini M, Fischbach A, Scharr H, Tsaftaris SA. Finely-grained annotated datasets for image-based plant phenotyping. Pattern Recognition Letters. 2016;**81**:80-89. [Internet]. Available from:. DOI: 10.1016/j.patrec.2015.10.013

Chapter 6

A Study on Traditional and CNN Based Computer Vision Sensors for Detection and Recognition of Road Signs with Realization for ADAS

Vinay M. Shivanna, Kuan-Chou Chen, Bo-Xun Wu and Jiun-In Guo

Abstract

The aim of this chapter is to provide an overview of how road signs can be detected and recognized to aid the ADAS applications and thus enhance the safety employing digital image processing and neural network based methods. The chapter also provides a comparison of these methods.

Keywords: Advanced Driver Assistance System (ADAS), digit recognition, digital image processing, neural networks, shape detection, road signs detection, road signs recognition

1. Introduction

Increasing population elevated the demand for personal vehicles and hence evolved the advancements in the vehicular designs, engine designs, and integration of embedded electronics making the personal vehicles one of the most integrated technologies of the everyday life [1, 2]. With personal vehicles becoming ubiquitous in everyday life, there has been a rise in the associated risks. As per the data from the U.S. Census Bureau, 10.8 million vehicular accidents have been recorded in the year 2009 compared to 11.5 million accidents in the year 1990 [3] marking the reduction in accidents by 6%.

With the evolution of progressive intelligence systems popularly referred as Advanced Driver Assistance System (ADAS) comprising of lane departure warning systems, forward collision warning system, road signs (speed limit and speed regulatory) detection and recognition system, driver drowsiness and behavioral detection and alert systems, and also adoption of passive safety measures such as airbags, antilock brakes, tire pressure monitoring systems or deflation detection systems, automated parking systems, infrared night vision, pre-crash safety system and so on have not only increased the driver safety but also resulted in the reduction of associated risks as these technologies continuously monitor the driver as well as their and vehicular environment and provides timely information and warnings to the driver.

IntechOpen

The detection and recognition of road signs is an important technology for the ADAS. Road signs are a guide to the drivers about the directions on the road, conditions of the road and serve as an essential warning under certain special road conditions. Thus, they enhance the road safety by providing the vital information. However, there might be the cases where a driver is distracted, is under stress of life, work or traffic, suffering lack of concentration or overwhelmed leading to overlook the road signs. Therefore, a system to monitor the road ahead of the vehicle, recognizing road signs and alerting about the vital conditions of the road would be an excellent assistance to the drivers. Pointedly, the road signs detection and recognition, which is the topic presented in this chapter cautions driver about the various road signs in a particular stretch of highways/roads enabling the drivers to drive within those limits, taking care of the road conditions and preventing from any over-speeding dangers.

The branch of computer science engineering that enables the machines which is the ADAS system in this case, to see, identify, interpret, and respond to the digital images and videos is termed as *Computer Vision*, abbreviated as *CV*. Until the boom of machine learning[1] techniques, CV was largely depended on traditional digital image processing (DIP)[2] methods, which are now mostly predicated on artificial neural networks (CNN)[3]. The impossible task for facilitating machines to respond to visions is achieved with the help of CV and it is intertwined with artificial intelligence[4].

The field of CV comprises of all tasks similar to biological vision systems such as seeing, i.e., visual senses, perceiving what is seen, draw detailed information in a pattern in which it can be used for further processes ultimately providing appropriate responses. In short, it is a modus operandi to instill humankind tendencies to a computer. CV finds its applications in the field of multiple disciplines aiding in simulating and automating the functions biological vision system employing sensors, computers, and various embedded platforms in assistance with numerous algorithms.

The applications of CV are enormous and broader. Of those numerous applications, using CV for detection and recognition of road signs to aid the Advanced Driver Assistance Systems (ADAS) is pivotal. This chapter focuses on road signs, also termed as traffic signs, detection and recognition using the key CV techniques.

The novelty of this chapter includes: (i) the proposed CV based method detects and recognizes the speed limit and speed regulatory signs without any templates as the templates are part of the code and not the images. (ii) the proposed CSPJacinto-SSD network enhances detection accuracy while reducing the model parameters and complexity compared to the original Jacinto-SSD.

---

[1] Machine learning (ML) is a branch of artificial intelligence (AI) and computer science, which focuses on the use of data and algorithms to imitate the way that humans learn, gradually improving its accuracy [4].

[2] Digital Image Processing (DIP) refers to the use of computer algorithms to perform image processing on digital or digitized images, leading to the extraction of attributes from the processed images and to the recognition and mapping of individual objects, features or patterns [5].

[3] An Artificial Neural Network (ANN) is a series of algorithms that endeavors to recognize underlying relationships in a set of data through a process that mimics the way the human brain operates [6].

[4] Artificial intelligence (AI) refers to the simulation of human intelligence in machines that are programmed to think like humans and mimic their actions. The term may also be applied to any machine that exhibits traits associated with a human mind such as learning and problem-solving [7].

2. Computer vision in ADAS applications

Computer Vision (CV) is one of the crucial technologies in building the smart and advanced vehicles with autonomous driving capabilities termed as Advanced Driving Assisting System (ADAS). One of the key arena of active research of the ADAS is the road signs detection and recognition, which is a challenging task. A number of issues such as the type of camcorder, the speed of a car, noises in the image depending on speed and direction, type and intensity of light and weather conditions and sometimes the background and other objects that are similar to the signs makes it monotonous to detect and recognize the road signs. Additionally, road signs may also be damaged, faded out, tilted and partially submerged by other objects such as building signboards, trees, leading to confusion in the automated system. The process of detecting the road signs of all types is carried out using the images/videos candidates comprising of targeted road signs in case of both DIP methods and CNN methods. The road signs can be obtained from various datasets such as German Traffic Signs Dataset (GTSDB) [8], Tsinghua-Tencent 100 K [9], ImageNet dataset [10], Pascal VOC [11] to name a few. Most of these vastly used datasets may not have all the road signs in sufficient numbers captured under different lighting and weather conditions. This leads to researches to build their own datasets or rely on mechanical simulations such as CarSim [12] to build the lacking traffic signs.

This chapter discusses a low-complexity DIP algorithm and CNN based method along with the existing researches, product embodiments of these technologies followed by the algorithm design, hardware implementation and performance results of road signs detection and recognition.

2.1 Road signs detection and recognition

The process of locating the road signs from a moving vehicle followed by recognizing the exact type of road signs can be termed as *'road signs detection and recognition.'* Although there are various approaches and different algorithms, some patterns may appear similar to that of an existing body of work as in **Figure 1** that shows the basic steps employed in road signs detection and recognition flow. The process is generally divided into three parts namely, road signs detection to locate the potential candidates of road signs followed by the verification of the detected road signs' candidates from the previous stages. Finally, the recognition of traffic signs to formulate the actual information from the detected and verified signs. This task of detecting followed by recognizing road signs to aid ADAS can be achieved through both DIP and CNN based methods.

Torresen et al. [13] presents a red-colored circular speed limit signs detection method to detect and recognize the speed limit signs of Norway. Moutarde et al. [14] presents a robust visual speed limit signs detection and recognition system for American and European speed limit signs. Keller et al. [15] presents a rectangular speed limit signs detection scheme aimed at detecting and recognizing the speed limit signs in United States of America (U.S.A.). A different approach is used by Liu et al. [16] wherein the de-noising method based on the histogram of oriented gradients (HOG) is applied to Fast Radial Symmetric Transform approach to detect the circular speed-limit signs. Zumra et al. [17] and Vavilin et al. [18] both uses color segmentation followed by other digital processing methods. Lipo et al. [19] presents the method that fusions camera and LIDAR data followed by the HOG and linear SVM to classify the traffic signs.

Sebastian et al. [20] presents the evaluation of the traffic signs detection in the real-world environments. The traffic signs are detected using the Viola-Jones detector

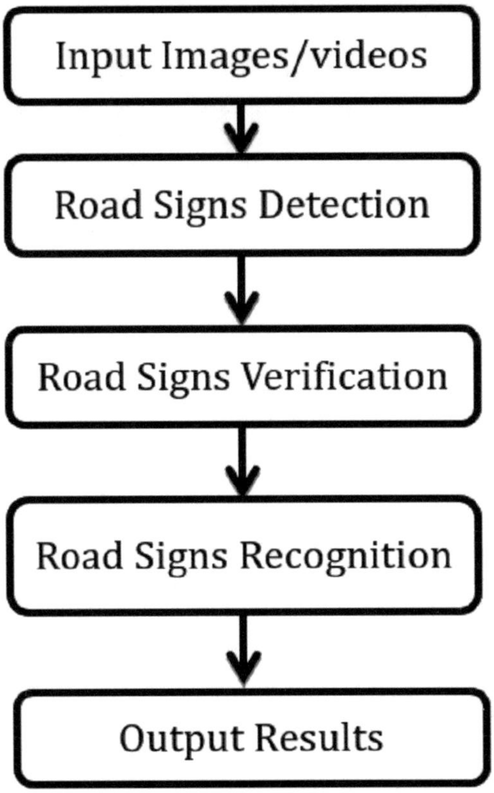

Figure 1.
Basic steps of road signs detection and recognition.

based on the Haar features and Histogram of Orientated Gradients (HOG) relied on linear classifiers. Model-based Hough-like voting methods are tested on the standard-The GTSDB. It also discusses different methods proposed by the Ming et al. [21] that uses two different, supervised modules for detection and recognition, respectively. Markus et al. [22] uses modern variants of HOG features for detection and sparse representations for classification and Gangyi et al. [23] presents the method that uses the HOG and a coarse-to-fine sliding window scheme for the detection and recognition of traffic signs, respectively.

Supreeth et al. [24] presents color and shape based detection scheme aimed at detection of red color traffic signs that are recognized using the auto associative neural networks. Nadra Ben et al. [25] presents a traffic sign detection and recognition scheme aimed at recognition and tracking of the prohibitory signs. Then Feature vector extraction along with the Support vector mechanism (SVM) is used to recognize the traffic signs and the recognized traffic signs are tracked by the optical-flow based method of Lucas-Kanade tracker [26]. Y. Chang et al. [27] adopted the modified radial symmetric transform to detect the rectangular patterns and then Haar-like feature based AdaBoost detector to reject the false positives. Abdelhamid Mammeri et al. [28] proposed an algorithm for the North American Speed limit signs detection and recognition. There are plenty of state-of-the-art researches based on different models of CNN [29–32] to detect and recognize the traffic signs including some hybrid approaches [33, 34].

2.1.1 Traditional digital image processing methods to detect and recognize road signs

The method employed to achieve certain operation on images with the aim of getting an enhanced image or extracting useful, interpretable information is termed as Image processing. It is similar to signal processing with the contraction that here input is an image and output is either an image or features affiliated with that image. In recent decades, image processing is among rapidly growing technologies. It forms the foundation for the computer vision and one of the core research area within engineering and computer science disciplines.

Fundamentally, image processing comprises of three steps namely, (i) Use of image acquisition tools to capture/import the images; (ii) Analyses and manipulation of the image; and, (iii) Output in which result can be altered image or report that is based on image analysis.

The methods used for image processing can be broadly classified in two, namely, analogue and digital image processing. Analogue image processing (AIP) refers to use of printouts and photographs to analyze via the basic interpretation employing visual techniques. On the other hand, digital image processing (DIP) techniques, as per the name, comprises of techniques that manipulate images digitally using computers. Pre-processing, enhancement, information extraction, and display are the basic, customary processes for all the data to undergo in DIP.

The process of detection and recognition of speed limit road signs [35] can be broadly classified into three stages namely, (i) speed limit signs detection, (ii) digit segmentation, (iii) digit recognition and that of detection and recognition of speed regulatory road signs [36] also into three stages such as, (i) speed regulatory signs detection, (ii) feature extraction, (iii) feature matching. **Figure 2** depicts the proposed algorithm used in detection and recognition of the road signs. The following sections discuss each step of the algorithm and the corresponding implementation specifications of the respective stages.

A. Shape Detection

The process of detecting regular and irregular polygon shapes is termed as *Shape detection*. Shape detection in this chapter refers to detecting the road signs which processes the entire frame and then focuses on selecting the potential candidates of size 32x32, 64x64 and so on comprising the common shapes, either a circle or a rectangle of the speed limit sign or a triangular signs of the speed-regulated signs using radial symmetric transform method.

The concept of radial symmetric transform [37, 38] uses the axes of radial symmetry. The normal polygenes of n-sides possess several axes of symmetry and the radial symmetric transform works based on these symmetric axes.

The voting process is based on the gradient of each pixel [39]. The direction of gradient generates a vote. The vote generated from each pixel follows the symmetric axes resulting in the highest votes at the center of the respective symmetric axes. **Figure 3** shows the radial symmetry for common polygenes.

Fundamentally, the Sobel operator [40] is applied to calculate the gradient of each pixel using a Sobel mask. Sobel operator generates the gradient of the intensities in the vector forms with the horizontal gradient denoted by G_x and vertical gradient denoted by G_y by convolving corresponding Sobel masks defining the direction of the gradients for each pixel. Besides, in order to eliminate the noises of small magnitudes, the

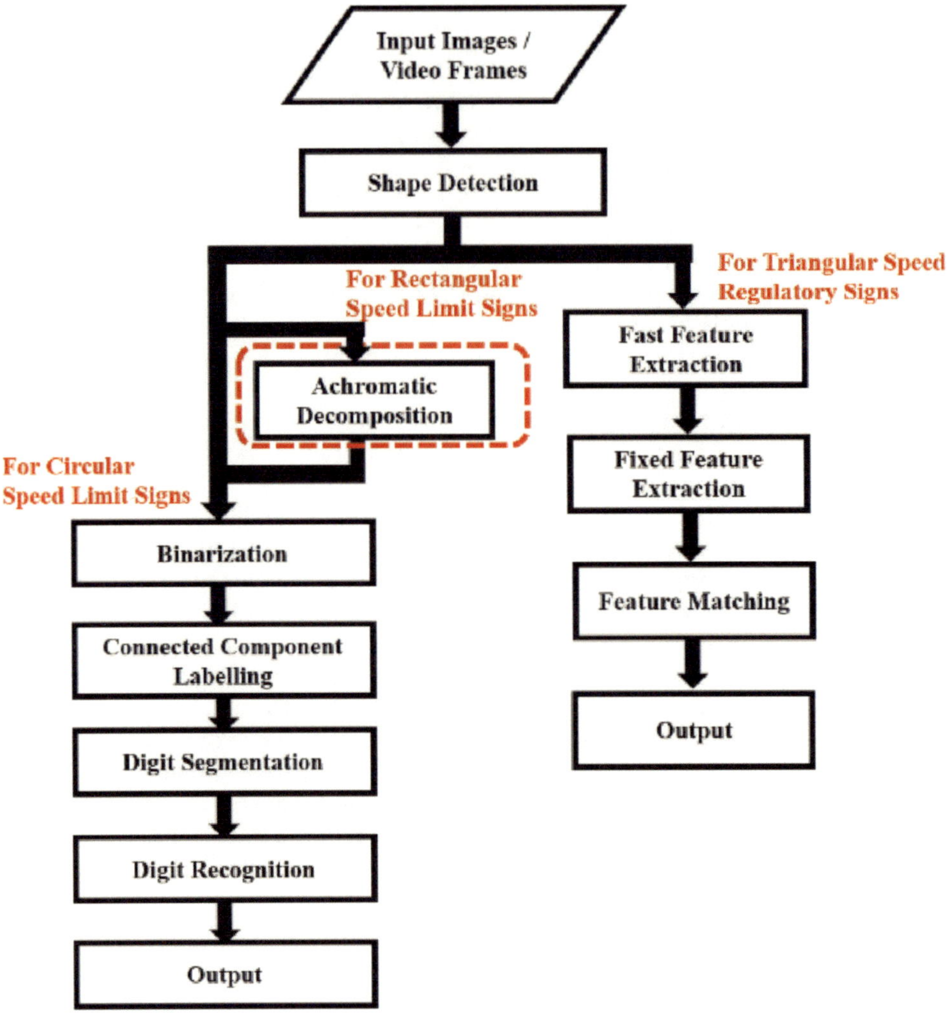

Figure 2.
Flow chart of the proposed algorithm used to detect speed limit and speed regulatory signs.

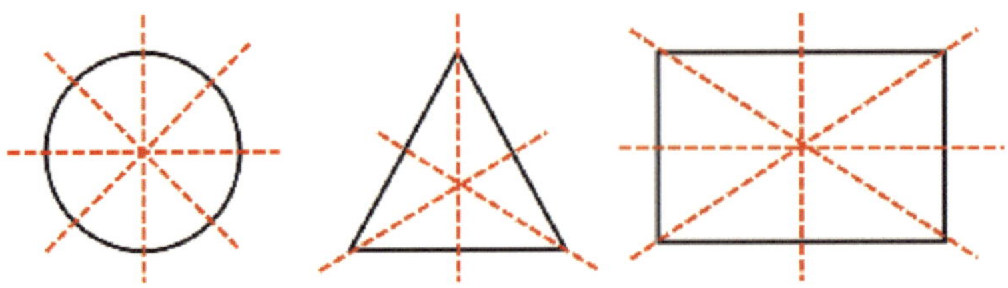

Figure 3.
Radial symmetry of common polygenes.

threshold for the absolute values G_{abs} is set for horizontal and vertical gradient given by Eq. (1)

$$G_{abs} = |G_x| + |G_y| \tag{1}$$

Once the horizontal gradient G_x and the vertical gradient G_y is obtained and the noise is eliminated, the radial symmetric transform can be processed based on the calculated gradients. **Figure 4** shows the results of the horizontal and vertical gradients.

i. Rectangular Radial Symmetric Transform

The voting process in the rectangular radial symmetric transform phase is based on the gradient generated from the Sobel operator [13, 27]. Each selected pixel with its absolute magnitude G_{abs} greater than a small threshold is denoted as p, and the gradient vector is denoted as $g(p)$. The direction of $g(p)$ can be formulated with the horizontal gradient G_x and the vertical gradient G_y into an angle using Eq. (2).

$$g(p) = \tan^{-1}\frac{G_y}{G_x} \tag{2}$$

For each considered pixel p, the votes along with the known width W and height H are divided into two categories- horizontal vote and the vertical vote. The direction of gradient $g(p)$ for each pixel is adopted to implement these two categories. The magnitude ranges and the ratio between G_x and G_y is used to verify the horizontal and the vertical votes with respect to the higher threshold and lower threshold values.

a. If $G_x >$ higher threshold and $G_y <$ lower threshold, vote is regarded as horizontal gradients.

b. If $G_x <$ lower threshold and $G_y >$ higher threshold, vote is regarded as vertical gradients.

Here, the values of higher threshold and lower threshold are experimentally chosen based on the size of the Sobel mask. In the case of 3x3 Sobel mask, the higher threshold is set in the range of 45–55, and the lower threshold is between 15 and 25. In case of the nighttime scenarios, both the thresholds are lowered to half of their original values and constrains are set for the ratio of horizontal and vertical gradients.

Each pixel contributes a positive vote and a negative vote. A voting line is then generated by each pixel with both positive and negative votes. The positive votes indicate the probable center of the speed limit sign while the negative votes indicate

Figure 4.
The results of the horizontal and vertical gradients.

the non-existence of speed limit signs. The positive horizontal votes $V_{horizontal+}$ and negative horizontal vote $V_{horizontal-}$ votes are formulated as in Eqs. (3) and (4). $L_{horizontal}\ (p,\ m)$ describes a line of pixels ahead and behind each pixel p at a distance W given by Eq. (5).

$$V_{horizontal+}\left\{L_{horizontal}(p,m)|m\ \epsilon\left[-\frac{W}{2},\frac{W}{2}\right]\right\} \tag{3}$$

$$V_{horrizontal-}\left\{\begin{array}{l} L_{horrizontal}(p,m)|m\ \epsilon\left[-W,-\frac{W}{2}\right) \\ \cup\left(\frac{W}{2},W\right] \end{array}\right\} \tag{4}$$

$$L_{horrizontal}(p,m)round(m*\overline{g}(p)+W*g(p)) \tag{5}$$

where $\overline{g}(p)$ is a unit vector perpendicular to $g(p)$. **Figure 5(a)** represents the process of horizontal voting. Similarly, the positive and negative vertical votes are formulated as in the Eqs. (6) and (7). $L_{vertical}\ (p,\ m)$ describes a line of pixels ahead and behind each pixel p at a distance W given by Eq. (8), and as shown in **Figure 5(b)** where $\overline{g}(p)$ is a unit vector perpendicular to $g(p)$.

$$V_{vertical}\left\{L_{vertical}(p,m)|m\epsilon\left[-\frac{H}{2},\frac{H}{2}\right]\right\} \tag{6}$$

$$V_{vertical}\left\{L_{vertical}(\mathrm{p},\mathrm{m})|\mathrm{m}\epsilon\left[-\mathrm{H},-\frac{\mathrm{H}}{2}\right)\cup\left(\frac{\mathrm{H}}{2},\mathrm{H}\right]\right\} \tag{7}$$

$$L_{vertical}(\mathrm{p},\mathrm{m})=\mathrm{p}+\mathrm{round}(\mathrm{m}*\overline{\mathrm{g}}(\mathrm{p})+\mathrm{W}*\mathrm{g}(\mathrm{p})) \tag{8}$$

After this voting process, the centers of the sign candidates will receive higher votes. The voting image is initially initialized to zero, and then it goes on accumulating both the positive and the negative votes. **Figure 6** shows the result for rectangular signs after the voting process.

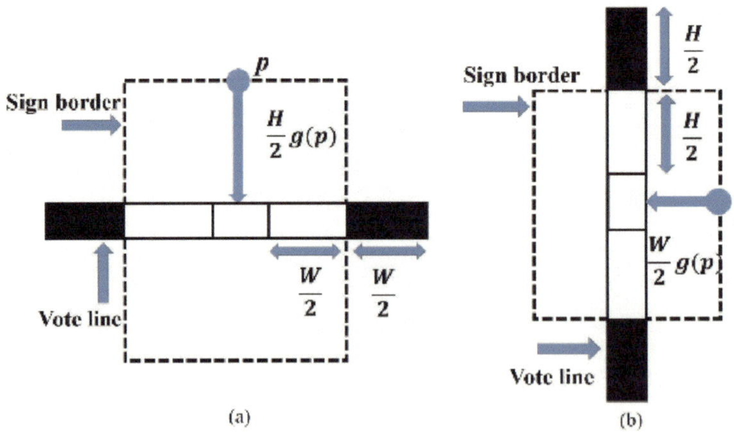

Figure 5.
(a) The voting line corresponding to the horizontal voting. (b) the voting line corresponding to the vertical voting.

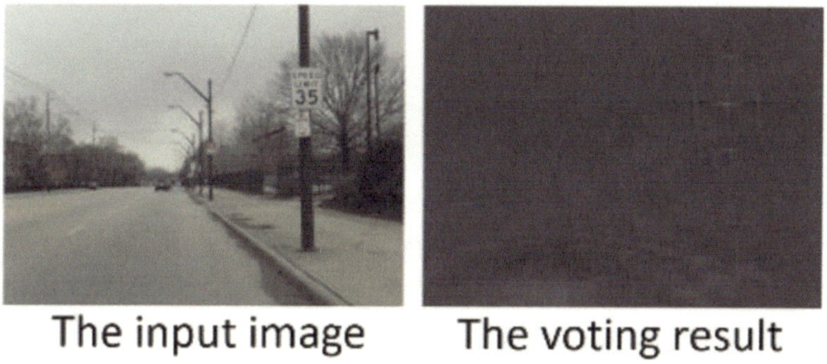

Figure 6.
The result after the horizontal voting process.

ii. Circular Radial Symmetric Transform

The detection of circular speed limit signs using radial symmetric transform method is similar to that of the detection of the rectangular signs with a difference that the circular radial symmetric transform need not to be divided into two parts as horizontal votes and vertical votes. It is entirely based on the direction of gradient for each pixel $g(p)$ and each considered pixel contributes only positive votes V_+ as in Eq. (9).

$$V_+ = \mathrm{p} + \mathrm{round}\ (\mathrm{R} * \mathrm{g(p)}) \tag{9}$$

Figure 7(a) illustrates the voting process for the circular sign detection and **Figure 7(b)** shows the result for circular signs after the voting process.

iii. Triangular Radial Symmetric Transform

The voting process of the triangular shape detection is also based on the gradient of each pixel [39]. The vote generated from each pixel follows the rule of the proposed triangle detection algorithm shown in **Figure 8**. It comprises of: (i) Sobel operator is applied to calculate gradient for each pixel. Consequently, we calculate the horizontal and vertical gradients by convolving the corresponding Sobel masks. Each selected pixel is represented with its absolute magnitude, and the gradient vector is denoted as $g(p)$. The direction of $g(p)$ can be formulated with the horizontal gradient G_x and the vertical gradient G_y into an angle as shown in Eq. (10). Only 180 degrees of gradient is used in this algorithm followed by the morphological erosion to eliminate noises.

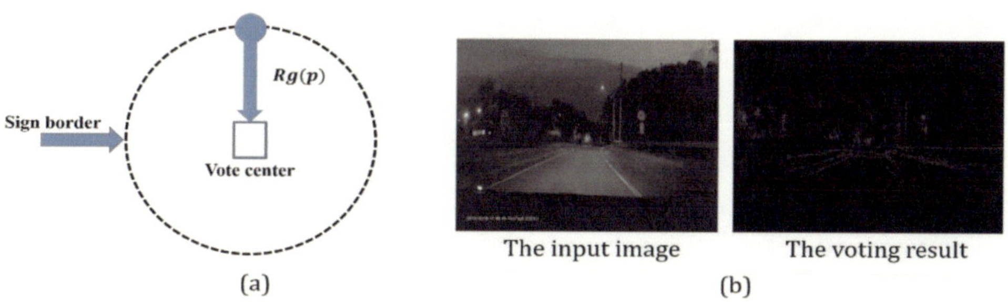

Figure 7.
(a) The vote center corresponding to (9); (b) the result of the circular voting result.

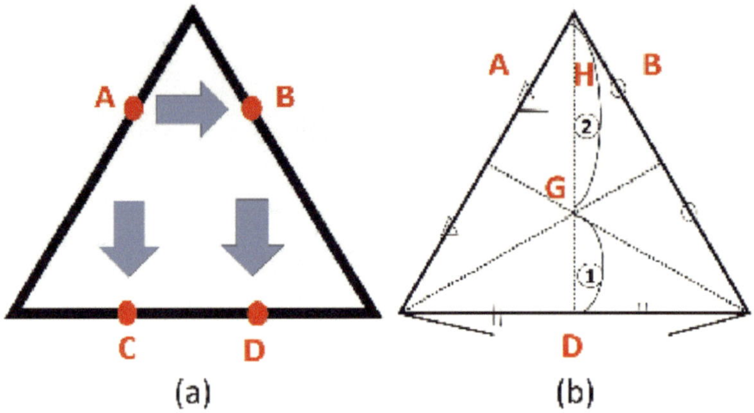

Figure 8.
(a) Illustration of triangle detection algorithm; (b) illustration of voting for the center.

Morphological erosion is applied to eliminate the noise. For a pixel $p(x,y,f(x,y))$ and a structure element $b(i,j)$, the formula of erosion is as in Eq. (11).

$$g(p) = tan^{-1}\frac{G_y}{G_x} \tag{10}$$

$$(f \ominus b)(x,y) = e(x,y) = \min_{(i,j)} (f(x+i, y+j) - b(i,j)) \tag{11}$$

Then the proposed algorithm exploits the nature of triangle for the detection as in **Figure 9(a)**. We look for the points having a gradient of 30 degree, defined as point A. Once the point A is obtained, we search for the points that have a gradient of 150 degrees on the same row as that of point A, defined as point B. The last step is to find the points C and D with 90 degrees gradient on the same column with points A and B, respectively. Once all these points are determined, a vote is placed to the point G at the centroid of the triangle as in **Figure 9(b)**.

In order to vote for the center point, the width of the detected position and the size of the target triangle is required to be calculated. The formulas shown in Eqs. (12) and (13), where $Center_x$ is the x-coordinate of G, $Center_y$ is the y-coordinate of G, H is the height of the small triangle, and D is the size of the target triangle, as shown in **Figure 9(b)**.

$$Center_x = \frac{A_x + B_x}{2}; H = (B_x - A_x) * \frac{\sqrt{3}}{2} \tag{12}$$

$$Center_y = \begin{cases} A_y + D * \frac{\sqrt{3}}{3} - H, H < D * \frac{\sqrt{3}}{3} \\ A_y + H - D * \frac{\sqrt{3}}{3}, H > D * \frac{\sqrt{3}}{3} \end{cases} \tag{13}$$

Then the width of the detected position is needed for calculating the detected points A and B. Thus, it is easy to build a look up table to reduce the computation cost

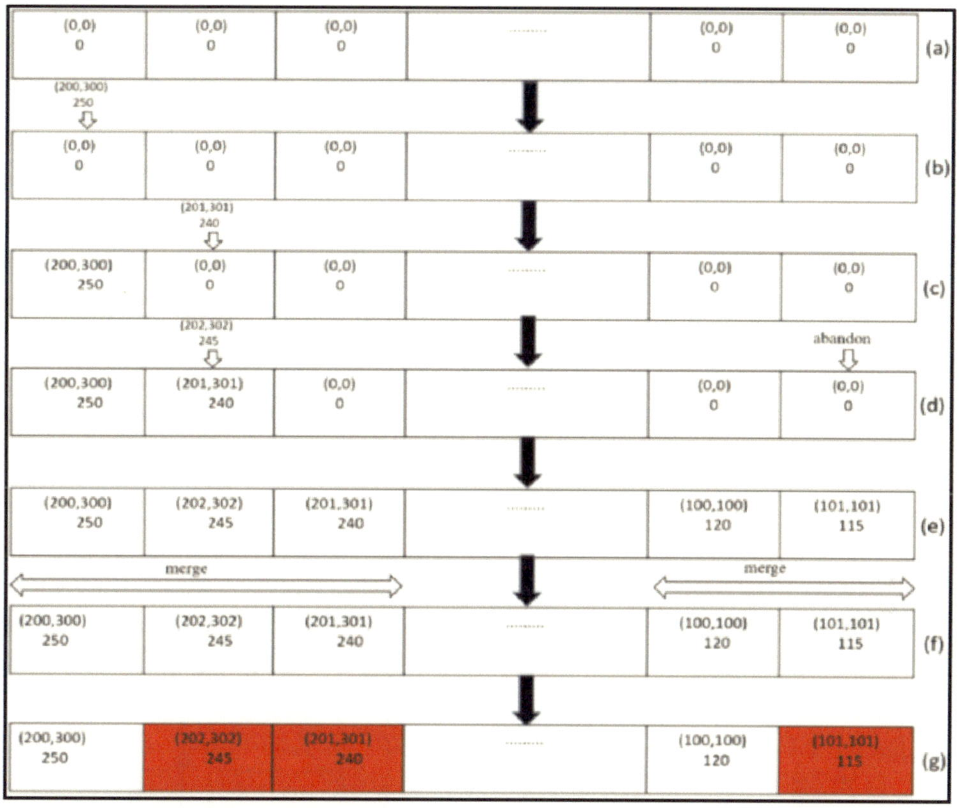

Figure 9.
The steps in detail of the sign candidate extraction (a) initialization of elements in the buffer to zero. (b), (c) insert the sign candidates based on the vote value. (d) if the current sign candidates own the greater vote value than any element in the buffer, firstly shift the element and the other elements in the wake of the right for the one-element and abandon the last element, and update the value to the element. (e), (f) post several iterations, the buffer is full of the sign candidates, and merge the cluster, only leaving the element with the greatest vote value. (g) the elements in red are merged.

of the voting process. The pixels with higher votes are judged to be in and near the center of the triangular road sign candidates. In order to reduce the computation cost, candidates that are close to each other will be merged into one candidate. The new coordinates of the candidate is the weighted arithmetic mean calculated using the coordinates of the merged candidates weighted by their vote thereby reducing the different candidates representing the same triangle.

B. Sign Candidates Extraction

After detecting the shape, the potential candidates of the road signs are extracted. A buffer is created to save all the potential sign candidates according to the following steps:

a. Initially, all the vote values in the buffer are set to zero.

b. For each vote in the input image, if the vote is greater than an experimentally set threshold, the pixel is considered. The considered vote is arranged into the buffer based on its vote value, ensuring the buffer is in a decreasing order. Every time this buffer is sorted in a decreasing order to ensure that, the pixels with greater values are in a prior order.

The values thus generated by the votes of the sign candidates result in a cluster of candidates in a small region. To overcome this challenge, the distance and search is defined from the prior order in the buffer by setting a small distance threshold to merge the cluster of sign candidates using non-maximum suppression as per Eq. (14). where x and y are the coordinates of the current considered sign candidates, and x_i and y_i are the sign candidates of the threshold candidates. **Figure 9** illustrates the details of the sign candidates' extraction along with the results of merging the cluster of sign candidates and the results of merging the clusters of the sign candidates.

$$Distance = |x - x_i| + |y - y_i| \tag{14}$$

C. Achromatic Decomposition

The key feature of the rectangular speed limit road signs in USA is that all the common speed limit signs are in gray-scale as in **Figure 10(a-d)**. There also exits advisory speed limit signs on the freeway exits as in **Figure 10(e-f)**. In order to detect the actual speed limit signs as in **Figure 11**, the achromatic gray scale color of the signs

Figure 10.
(a-d) the rectangular speed limits road signs in USA. (e-f) the advisory speed signs on freeway exits.

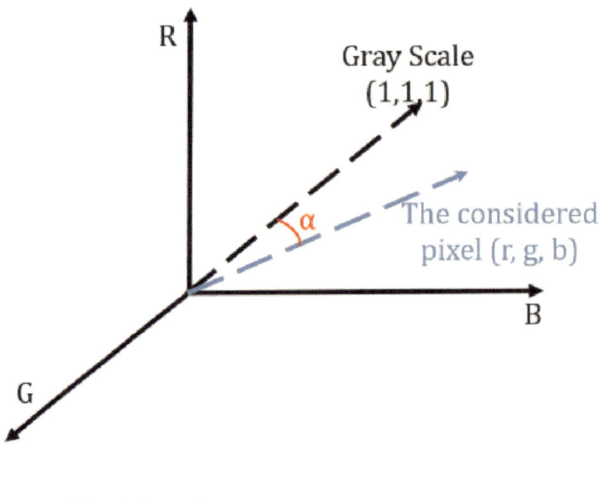

Figure 11.
The schematic of the RGB model and the angle α.

is extracted by the achromatic decomposition whereas the non-gray scale advisory speed signs are ignored from the further consideration.

The vector of gray scale is along (1,1,1) in RGB color space and the inner product is between (1,1,1) and the each considered pixel checks an angle α between these two vectors to apply the decomposition in RGB domain [41] as illustrated in **Figure 10(a-d)**.

Each of the considered pixel is in the vector form of (r, g, b). The cosine function of α, which is equal to the inner product [26] is shown in Eq. (15).

$$\mathrm{Cos}\alpha = (1,1,1)\cdot(r,g,b) = \frac{r+g+b}{\sqrt{3}\times\sqrt{r^2+g^2+b^2}} \tag{15}$$

For the implementation in our proposed, a mere value of cos^2 is considered. If the value is near to one, α is near to zero, which implies the considered pixel is in gray-scale and is taken in account for the further steps. **Figure 12** shows the results of the achromatic decomposition where the speed warning signs of non-gray scale found on the freeway exits are not acknowledged.

D. Binarization

In the proposed system, the Otsu threshold method is used for the daylight binarization while the adaptive threshold method during the nighttime. **Figure 13** illustrates these proposed steps.

To differentiate between daylight and the night-light, the ROI is set to the cent er part of the frame choosing the width ROI_w and height ROI_h as in Eqs. (16) and (17) where the sky often lies. The noise with extremely high and low pixel values are filtered out and then in the remaining 75% of the pixels, the average of pixel values is calculated to judge if it is a day-light or a night condition. **Figure 14** shows the schematic of the day and nighttime judgment.

$$ROI_{\mathrm{h}} = \frac{Height\ of\ the\ frame}{6} \tag{16}$$

$$ROI = \frac{2}{3}Width\ of\ the\ frame;\ Excluding\ \frac{Width\ of\ the\ frame}{6}\ on\ either\ ends \tag{17}$$

The Otsu method [42] can automatically decide the best threshold to binarize well in daytime, but at night, the chosen threshold causes the breakage of the sign digit.

Figure 12.
The results of the achromatic decomposition.

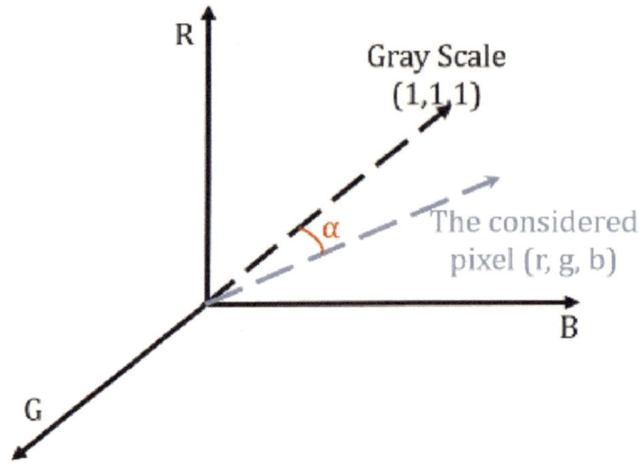

Figure 13.
The proposed steps of binarization.

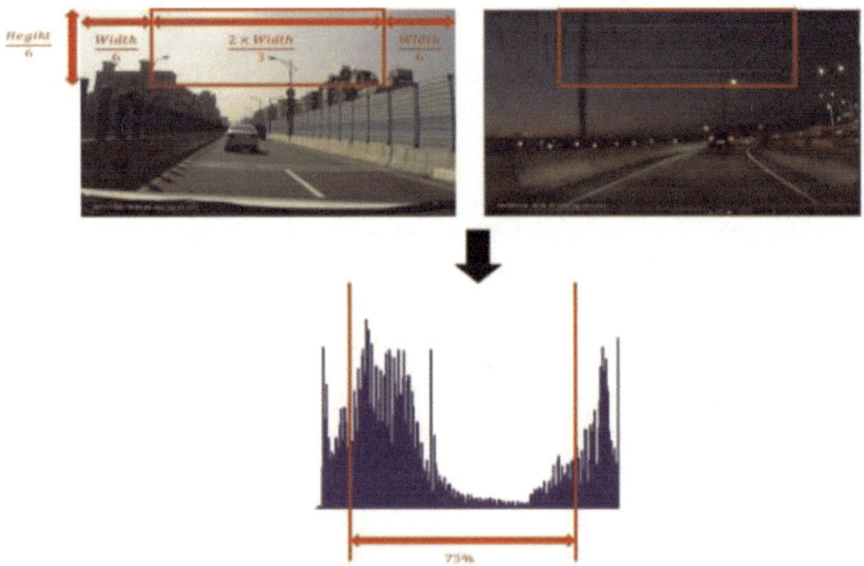

Figure 14.
The schematic of the day and night judgment.

In Otsu method, the best threshold that can divide the content into two groups and minimize the sum of variances in each group is found by an iterative process. **Figure 15** shows the schematic steps of Otsu method. On the other hand, the adaptive threshold is more sensitive. It divides the signs into several sub-blocks, mean of each block is calculated, and then the threshold for respective sub-blocks is computed based on the means in each sub-block. The corresponding results of the adaptive threshold is as shown in **Figure 16**.

Therefore, we chose Otsu threshold to automatically select the best-fit threshold in automatically daytime and adaptive threshold at night to handle the low contrast environment.

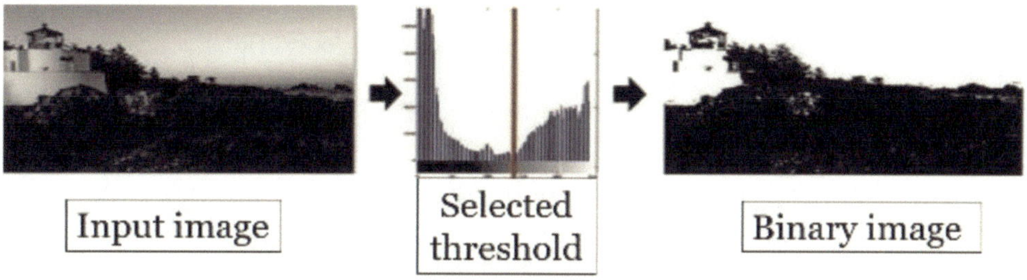

Figure 15.
The schematic steps of Otsu method.

Figure 16.
Different thresholding results.

For acceleration, this paper adopts the integral image [36] in which each pixel is compared to an average of the surrounding pixels. An approximate moving average of the last s-pixels seen is specifically calculated while traversing the image. If the value of the current pixel is lower than the average then it is set to black, otherwise it is set to white. In the proposed algorithm, it is considered to be stored at each location, $I(x, y)$, the sum of all $f(x, y)$ terms to the left and above the pixel (x, y). This is accomplished in linear time using Eq. (18) for each pixel. Once the integral image is first calculated, the sum of the function for any rectangle with the upper left corner (x_1, y_1), and lower right corner (x_2, y_2) can be computed in constant time using Eq. (19). The schematic of Eq. (13) can be illustrated with **Figure 17** and Eq. (19) can be modified into Eq. (20). Finally, the mean of each sub-block can be calculated and then

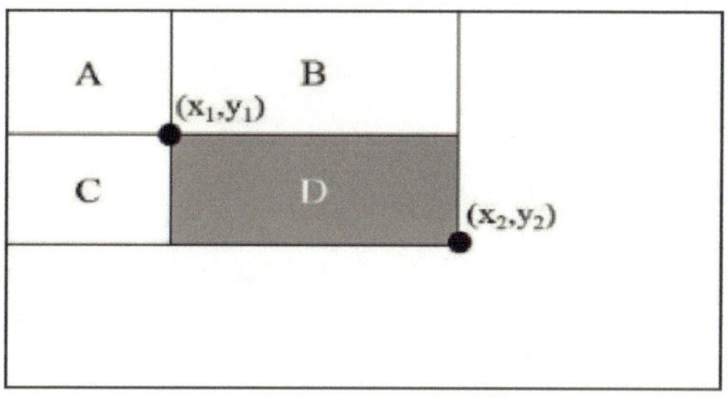

Figure 17.
The schematic figure of Eq. (18).

each pixel in the sub-block which in terms of $A(x, y)$ can be binarized with Eq. (21) where $\propto$ is a scalar based on the contrast under different conditions.

$$I(x,y) = f(x,y) + I((x-1),y) + I(x,(y-1)) - I((x-1),(y-1)) \tag{18}$$

$$\sum_{x=x_1}^{x_2}\sum_{y=y_1}^{y_2} f(x,y) = I(x_2,y_2) - I(x_2,y_1-1) - I(x_1-1,y_2) + I(x_1-1,y_1-1) \tag{19}$$

$$\mathrm{D} = (\mathrm{A}+\mathrm{B}+\mathrm{C}+\mathrm{D}) - (\mathrm{A}+\mathrm{B}) - (\mathrm{A}+\mathrm{C}) + \mathrm{A} \tag{20}$$

$$A(x,y) = \begin{cases} 255, & \text{if } I(x,y) > T_{avg} \times \propto \\ 0, & \text{otherwise} \end{cases} \tag{21}$$

E. Connected Component Labelling (CCL)

Connected component labelling (CCL) labels the object inside the sign candidates with the height, width, area and coordinate information. The CCL algorithm [28] is divided into two processing passes namely first pass and the second pass. Two different actions are taken in these passes if the pixel iterated is not the background.

The steps of connected component labeling are illustrated in **Figure 18**. In this case, the equivalent labels are (1, 2), (3, 7) and (4, 6).

F. Digit Segmentation

The purpose for digit segmentation [43] is to extract the digit from the binarized image. In the process of rectangular speed limit signs detection, the signboards have the characters reading "SPEED LIMIT" alongside the digits. As a result, it is necessary to set constrains on size of the digit candidates as per Eqs. (22) and (23). Similarly, the constraints on the size of the digit candidates in circular speed limit signs are as per Eqs. (24) and (25).

$$0.15 \text{ x } W \leq \text{Digit width} \leq 0.5 \text{ x } W \tag{22}$$

$$0.15 \text{ x } H \leq \text{Digit height} \leq 0.5 \text{ x } H \tag{23}$$

$$0.125 \text{ x } R \leq \text{Digit width} \leq R \tag{24}$$

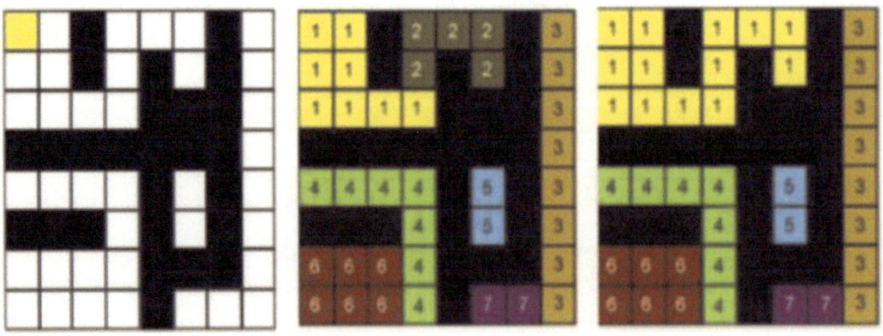

Figure 18.
The steps for connected-component labeling, (a) processing initialization, (b) the result after the first pass, (c) the result after the second pass.

$$0.5 \text{ x } R \leq \text{Digit height} \leq 1.5 \text{ x } H \tag{25}$$

Considering the fact that rectangular speed limit signs consist two-digits alongside the characters, it must be ensured that the selected candidates are of digits of speed limit sign and not the characters. The pairing rule of sizes and positions proposed in this paper are as follows:

a. The areas of the digit candidates should be similar.

b. The positons of the digit candidates should be closely packed.

c. The density of the pixels inside the digit candidates should be similar.

Whereas in the circular speed limit detection that are of both 2-digits and 3-digits, a loose constrain is adopted as it has only digits inside the circular speed limit signs and no characters. The pairing steps are similar to those followed in the rectangular speed limit signs. **Figures 19** and **20** shows the segmentation results of the rectangular and circular speed limit signs, respectively.

There exists a critical challenge in binarization process, as the digits may appear connected to each other. To overcome this challenge, a two-pass segmentation process is proposed in this paper. Digit segmentation, similar to the previous segmentation

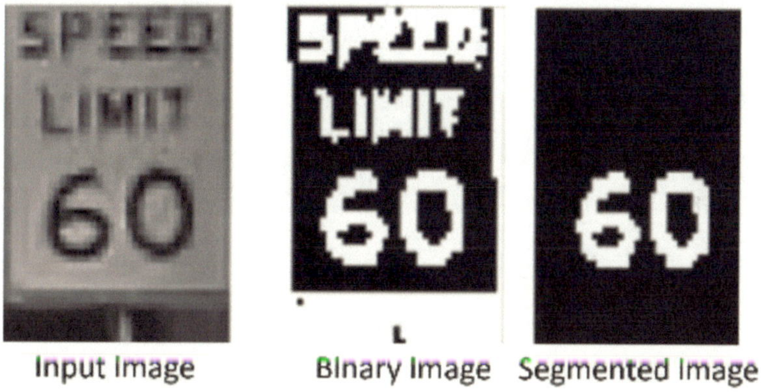

Figure 19.
The example of digit segmentation results of rectangular speed limit signs

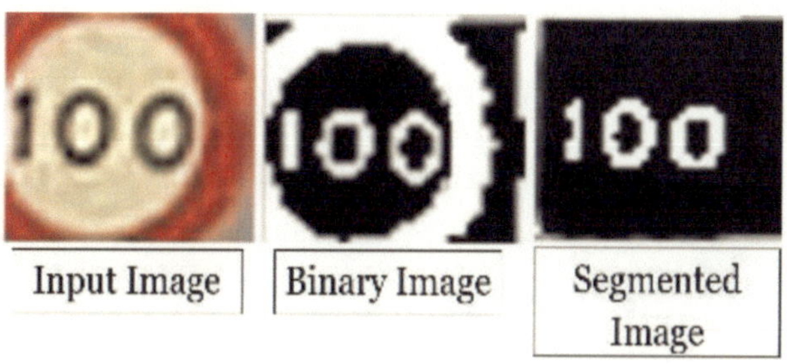

Figure 20.
The example of digit segmentation results of circular speed limit signs.

process, is applied. If large components are detected, then the second pass of the segmentation is applied.

The horizontal pixel projection is applied in the second pass segmentation. During this projection, the total number of pixels in each column is accumulated choosing the segment line based on the horizontal projection. **Figure 21** shows the example of horizontal projection and the result of the two-pass segment step.

G. Digit Recognition

In the digit recognition phase, the extracted digits are compared with the built-in templates and three probable digits with least matching difference are selected based on Sum of Absolute Difference (SAD) [44]. After which, the blob and breach features of the digits are applied to verify the final digits [45–47]. **Figure 22** depicts the proposed steps for digit recognition.

After selecting the three probable digit candidates, the blob feature is employed to verify the digit. Here the blob is defined as a close region inside the digit and is detected and gathered in several rows forming a union row. The pixel value of a union row is the union of values of the rows in that union row. For each union row, the number of lines in the white pixels are counted. A blob is formed only if the number of lines is of the sequence “1, 2,. ... , 2, 1”. This union row method yields the position and

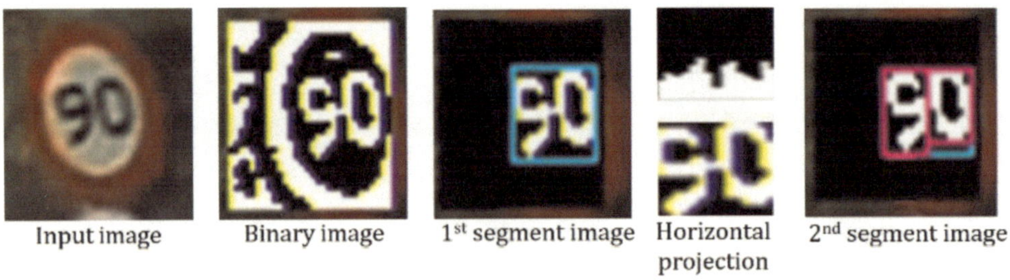

Figure 21.
Example of horizontal projection and the result of the proposed two-pass segmentation steps.

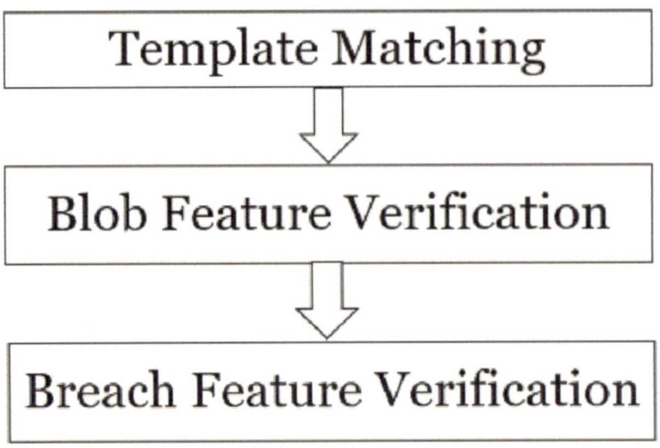

Figure 22.
The proposed steps for digit recognition.

Digit	Blob feature verification		Breach feature verification	
	Number of blobs	**Position of blobs**	**Number of breaches**	**Direction of breaches**
0	One	Top to Bottom	Zero	—
1	Zero	—	Zero	None
2	Zero	—	Two	Top Left & Bottom Right
3	Zero	—	One	Left
4	One	Top Left	Zero	—
5	Zero	—	Two	Top Right & Bottom Left
6	One	Lower Half	One	Top Right
7	Zero	—	One	Left
8	Two	Upper & Lower Halves	Zero	—
9	One	Upper Half	One	Bottom Left

Table 1.
The blob and breach feature verification for all the digits.

the number of the blobs in the digit candidate as in **Table 1**. The exact blob feature is defined for the specific digits.

Similarly, the breach feature is also adopted to verify the digits. A breach is defined as an open region formed by a close region with a gap. The breach is detected by counting the number of pixels where the white pixel first appears from both the right and the left in each column to half the width of the digit candidates. If there is a series of pixels that are larger than half the digit height, then it is regarded as a breach. **Table 1** shows the blob and breach feature verification for the digits from 0 to 9 and **Figure 23** shows the results of the digit recognition in terms of blob and breach feature verification.

H. FAST Feature Extraction

Features from Accelerated Segment Test (FAST) [48, 49] is a high repeatability corner detector. As shown in **Figure 24**, it uses a circle of 16 pixels to classify whether or not a candidate point p is a corner. The FAST feature extraction conditions can be written as in Eq. (26) where S is a set of N contiguous pixels in the circle, I_x is intensity of x, I_p is intensity of candidate p and t is the threshold.

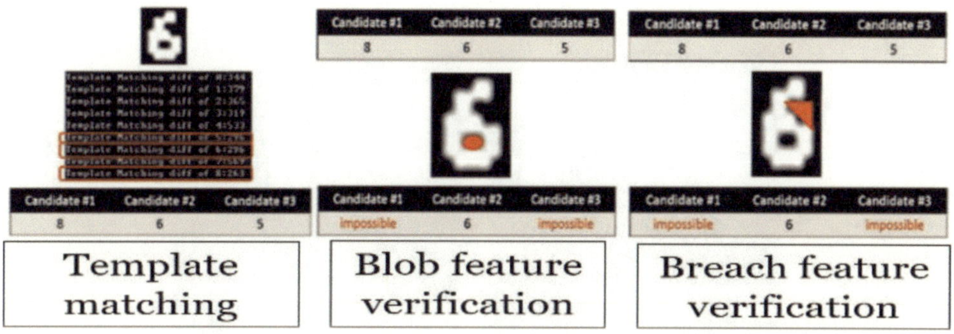

Figure 23.
Digit recognition results.

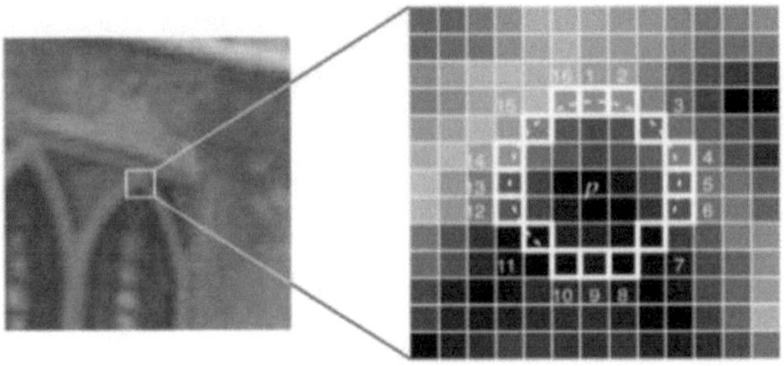

Figure 24.
The illustration of FAST algorithm.

$$\forall x \in S, I_x > I_p + t, \text{and } \forall x \in S, I_x < I_p - t \tag{26}$$

There are two parameters to be chosen in FAST algorithm namely, the number of contiguous pixels N and the threshold t. N is fixed as 9 in the proposed algorithm whereas to overcome the changes in the intensity inclinations as shown in **Figure 25**, the threshold t is set to be dynamic. The dynamic threshold is calculated by the image patch of the sign candidates. First, we count for pixels with intensity bigger than 128. If the number of bright pixels is between 20% and 80% of the total number of pixels in the image patch, the threshold is computed by the percentage of number of bright pixels over the total number of pixels. There are two fixed thresholds for the conditions that the number of bright pixels is lower than 20% of the total number of pixels or higher than 80% of the total number of pixels. Accordingly, the threshold dynamically updates to the number of bright pixels over the total number of pixels.

I. Fixed Feature Extraction

There are certain conditions in which the contents of road signs are too simple to be extracted using the FAST feature, as shown in **Figure 26(a-c)**. Thus, the Fixed Feature Extraction is applied to handle these road signs with good inclinations in the proposed algorithm.

Fixed Feature Extraction uses thirty fixed feature points to describe a road sign, as shown in **Figure 26(d)**. This method is similar to template matching, but it is more robust to noises as it uses descriptors to describe the small area around the feature points.

Figure 25.
Different lighting conditions of road signs.

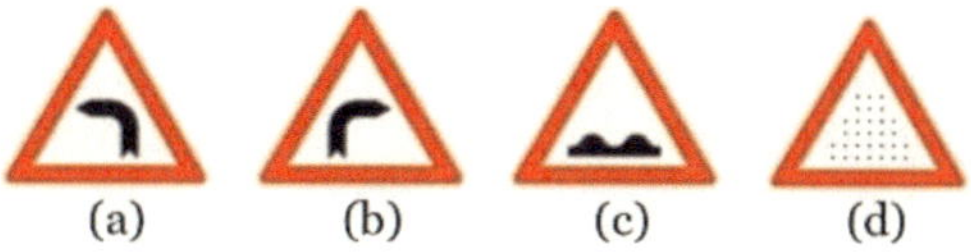

Figure 26.
(a-c) Road signs with simple contents; (d) thirty fixed feature points used in fixed feature extraction.

J. Feature Matching

The main objective of this phase is to match the features between the pre-built template and the detected sign candidates as shown in **Figure 27**. The features extracted previously are matched by their coordinates and the descriptors which are constructed to describe their respective features. Due to the fact that the proposed system is aimed at real-time applications, the construction and the matching procedure of the descriptor algorithm should be both simple and efficient at the same time.

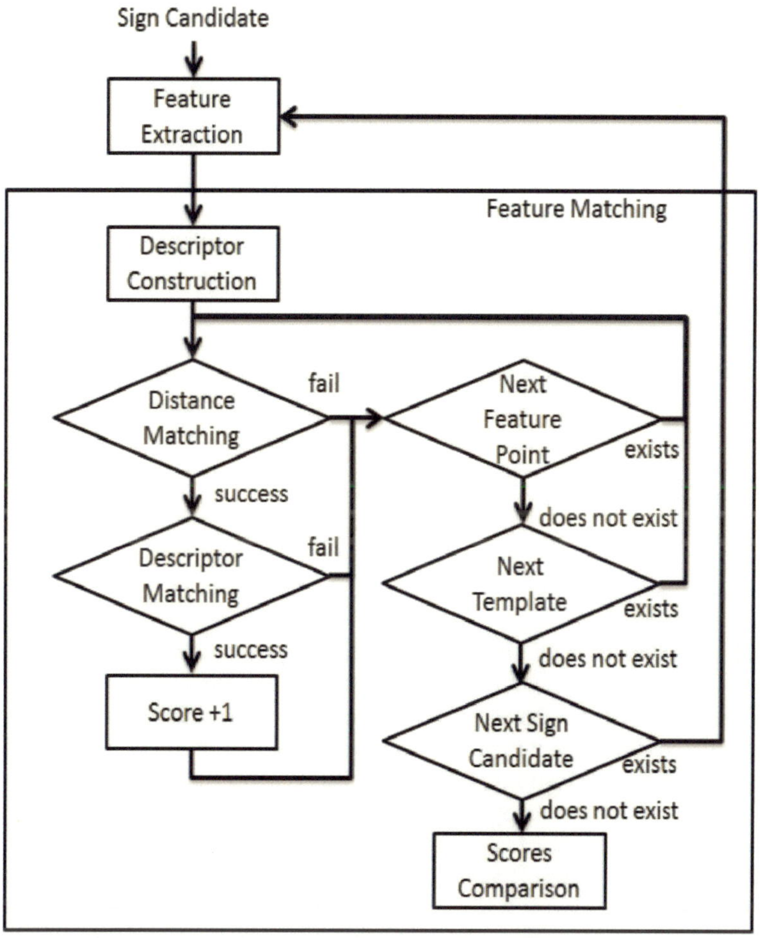

Figure 27.
Steps followed in the feature matching.

Binary Robust Independent Elementary Features (BRIEF) [50] is a simple descriptor with good matching performance and low computation cost. In order to build a BRIEF descriptor of length n, n pairs (x_i, y_i) are chosen. X and Y representing the vectors of point x_i and y_i respectively, are randomly sampled with the Gaussian distribution and stored in a pre-built array to reduce the computation cost. They are sampled with Gaussian distribution stored in a pre-built array to reduce computation cost. To build a BRIEF descriptor, τ test is defined as in Eq. (27) and n is chosen as 256 to yield the best performance.

$$\tau(p; x, y) := \begin{cases} 1, p(x) < p(y) \\ 0, p(x) \geq p(y) \end{cases} \tag{27}$$

The advantages of BRIEF are obviously, low computation time and better matching performance whereas the disadvantage is that the BRIEF is not rotation invariant and scale invariant. Since the size of the detected signs is fixed and the road signs would not have too much rotation effect, these disadvantages do not influence the recognition result.

After descriptor construction, a two-step matching process comprising distance matching and descriptor matching is applied to match the detected sign candidates with the pre-built templates. Distance matching considers only the coordinates of the feature points. In this application of road signs recognition, the detected road signs should be a regular triangle sometimes with certain defects such as, lighting changes, slight rotation, and occluded with an object. Thus, the two similar feature points are not matched if the coordinates of these two feature points are different.

The goal of descriptor matching is to compute the distance between two descriptors, one is from the detected sign candidate and the other is from the pre-built template. As with all the binary descriptors, the measure of BRIEF distance is the number of different bits between two binary strings which can also be computed as the sum of the XOR operation between the strings.

After all sign candidates are matched, the scores comparison is applied to choose which template is the most suitable for final recognition result. The template with the scores higher than the others is judged as the result of the template matching. Moreover, the same result should be recognized a few times in several frames of a video to make sure that the result does not produce a false alarm.

The performance of the aforementioned DIP based algorithms in detecting and recognizing the road signs are discussed in the Section 3.

2.1.2 Computer neural network (CNN) methods to detect and recognize road signs

Artificial Neural networks (ANN) generally referred as Neural networks (NN), specifically as Computer Neural Networks (CNN) have been a sensation in the field of CV. ANNs, Artificial Intelligence (AI) and Deep Learning (DL) are interdependent and importantly, indispensable topics of recent years research and applications in engineering and in the technology industry. This reason for this prominence is that they currently provide the best solutions to many problems extensively in image recognition, speech recognition and natural language processing (NLP).

The inventor of one of the first neurocomputers, Dr. Robert Hecht-Nielsen defines a neural network as "*...a computing system made up of a number of simple, highly interconnected processing elements, which process information by their dynamic state response to external inputs.*" In simpler words, the theme of ANNs are motivated by biological neural networks to learn and process the information fed to it. **Figure 28**

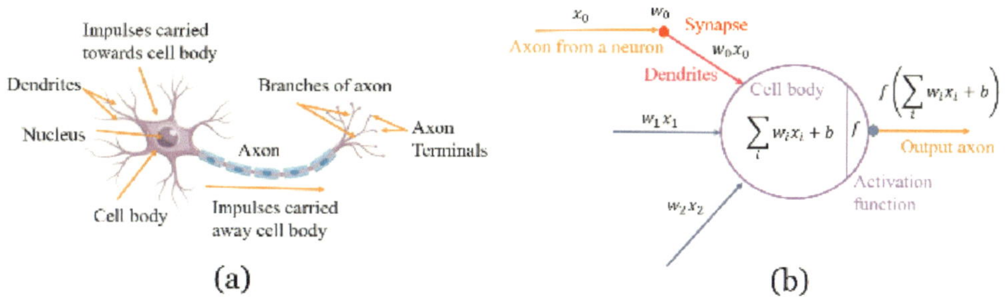

Figure 28.
(a) A representative biological neuron with (b) its mathematical model from [34].

shows the similarity in function of a biological neuron in **Figure 28(a)** with its respective mathematical model in **Figure 28(b)**.

A neuron is the fundamental unit of computation in a biological neural network whereas the basic unit of an ANN is called a node or a unit. The node/unit receives inputs from external sources and from other nodes within the ANN to process and computes an output. Every input has a characteristic weight (w) allotted based on its corresponding importance to other inputs. The node applies a function to the weighted sum of its inputs.

ANNs are generally organized in layers that are made up of numerous interconnected 'nodes' comprising an 'activation function' as in **Figure 29**. The inputs are presented to the ANN via the 'input layers', which communicates with one or more 'hidden layers' in which a particular processing is done by a system of weighted 'connections'. The hidden layers then link to an 'output layer' where the answer is output. For the general model of ANN in **Figure 29**, the net input can be calculated as in Eq. (28) and the output by applying the activation function over the net input can be calculated using Eq. (29).

$$Y_{in} = X_1.w_1 + X_3.w_2 + \ ...X_n.w_n \ i.e., \text{the net input}, Y_{in} = \left(\sum i - n\right)X_i.w_i \quad (28)$$

$$Y = F(Y_{in}) \quad (29)$$

Recent researches and publications present that ANNs are extensively used for various applications ranging from object detections to learning to paint, to create images from the sentences, to play board games-AlphaGo and so on. There are many more unthinkable and unconvincing things done by the ANNs in the present days and research studies on further advancing them are going on rigorously.

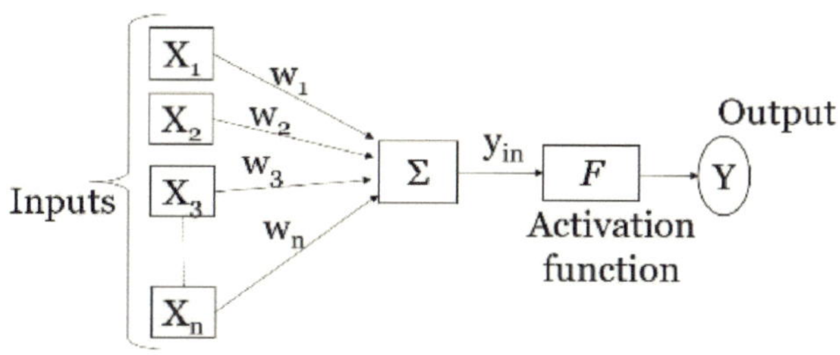

Figure 29.
The general model of an ANN.

There exist a various models presented from researches across the word for different applications. Some of the prominent CNNs are Single Shot Detector (SSD) [51], Region based Connect Neural Network (R-CNN) [52], Fast-RCNN [53], Faster-RCNN [54], You Only Look Once (YOLO) [55] and different versions of it, Generative Adversarial Networks (GAN) [56] and different modules [57] based on it and many more. This chapter also discusses CSPJacinto-SSD based on CSPNet [58] features in JacintoNet [59]. These innumerous ANNs are extensively used by the researchers and industries alike. The researchers and industries go hand-in-hand to investigate on further improvisations of the existing NNs, expanding them into diversified applications and solving a problem/challenge using effective and low-cost measures, ultimately manufacturing commercial products to make human lives easier and smarter.

In this chapter, we explore object detection NNs such as SSD, Faster-RCNN, YOLO and propose the newer CNN model termed as 'CSPJacinto-SSD' for the detection and recognition of road signs.

The SSD, as its name suggests, only need to take one single shot to detect multiple objects within the image. It has two components- a backbone and SSD head. The backbone model is a network used for pre-trained image classification. The SSD head is network with one or more convolutional layers added to the backbone network and the outputs are interpreted as the bounding boxes and classes of objects in the spatial location of the final layers activations as in **Figure 30**.

Faster-RCNN [54] comprises of two modules of which the first module is a deep fully convolutional network that proposes regions, and the second module is the Fast R-CNN detector [53] that uses the proposed regions. The earlier version of the Faster-RCNN such as R-CNN and Fast R-CNN both use selective search method to find out the region proposals. The selective search method is a slow, hence time-consuming process affecting the performance of the network. To overcome these challenges, an advanced version of the R-CNN called Faster-RCNN was proposed [54] that has an object detection algorithm eliminating the selective search algorithm and the network learn the region proposals. **Figure 31** shows a Faster-RCNN network.

All of the previous object detection algorithms until the year 2015, used regions to locate the objects in the image. That means the network does not look at the complete image but only parts of the image that may the high probabilities of containing an object. In 2015, J. Redmon proposed a new NN called YOLO (You Only Look Once) [55] as in **Figure 32**. It is an object detection algorithm much different from the region-based algorithms. In YOLO, a single convolutional network predicts the bounding boxes and the class probabilities for these boxes.

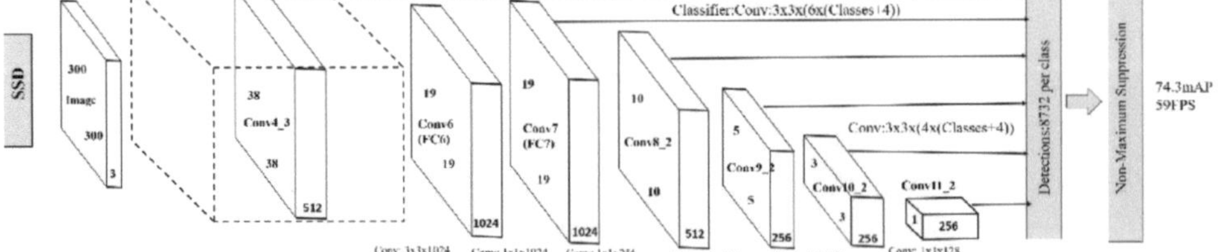

Figure 30.
SSD model adds several feature layers to the end of a base network, which predict the offsets to default boxes of different scales and aspect ratios and their associated confidences.

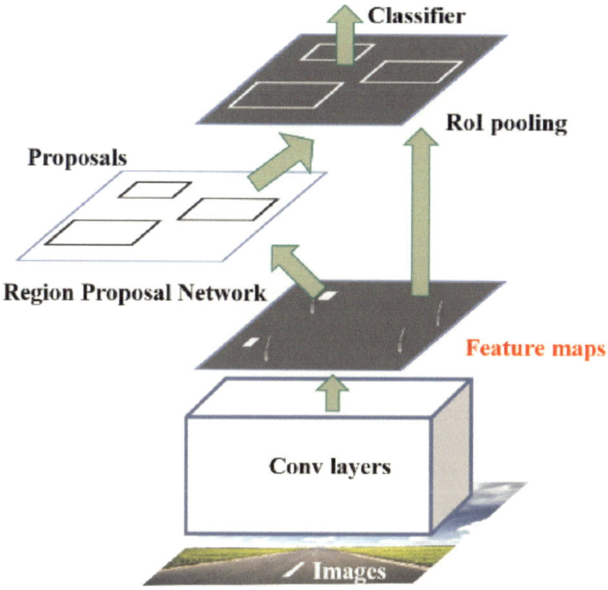

Figure 31.
A single, unified faster R-CNN for object detection.

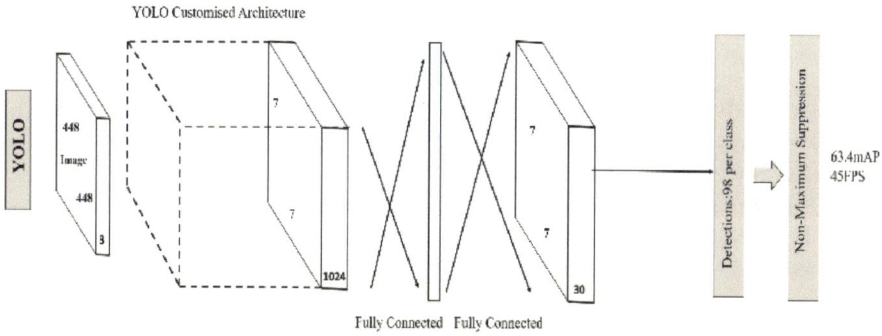

Figure 32.
A representative of the YOLO architecture for object detection.

The overall architecture of CSPJacinto-SSD is shown in the **Figure 33**. The CSPNet [58] features are added in JacintoNet [59], which is a simple light-weighted model composed of convolution, group convolution, and max-pooling layers. The Cross Stage Partial (CSP) feature is proved to improve the accuracy while reducing the model parameters and complexity. The function of CSP is simply to split the feature maps into two parts along channels at the input of each stage, one part sends into the convolution block, as usual, the other part skips all layers and concatenate with the output convolution block together as the final block output. In **Figure 33**, one blue and one green square can be seen as a convolution block. The blue arrows show the CSP feature as described above, and the red arrows show the output of each stage. The 1 x 1 convolution before the convolution block is used to increase the feature channels, and the 1 x 1 convolution after the convolution block is used to merge the context of features from the CSP layer. Out1 to Out5 labels the feature maps that are used for dense heads to process the bounding box outputs.

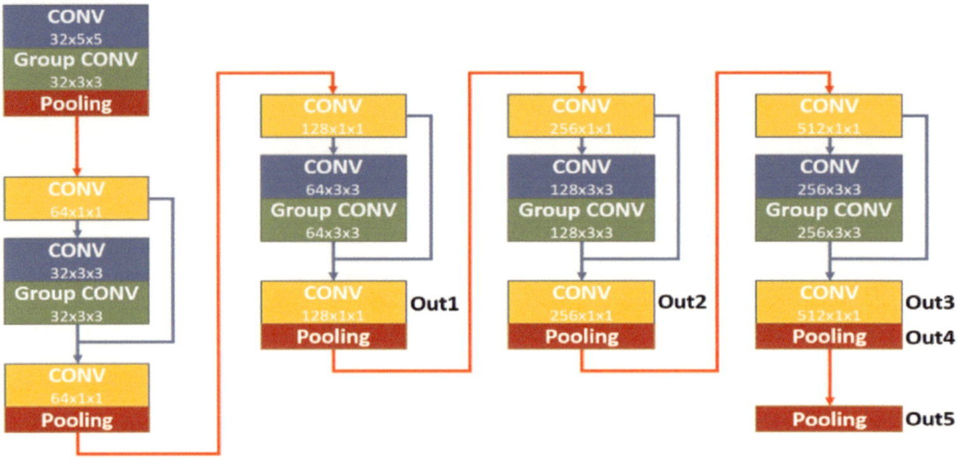

Figure 33.
CSPJacinto-SSD model architecture.

	Base sizes
Original SSD	16, 32, 64, 100, 300
Proposed CSPJacinto-SSD	16, 32, 64, 128, 256

Table 2.
Base size of anchor box.

The dense heads employed in the proposed CSPJacinto-SSD are referred to those in SSD, with some modifications in the anchor boxes based on the concept of multi-head SSD proposed in [60]. At dense head levels 2 to 4, there is an extra location of anchor boxes with offset 0, instead of only the original offset 0.5. This feature can increase the anchor boxes density improving the recall of object detection, especially used for light-weighted SSD models that need more anchor boxes to guide the objects' possibly appeared location.

The anchor box settings are a bit different from the original SSD model. The anchors 1:2 is changed to 1:1.5 because it may make the anchor borders denser, and preserve 1:3 anchors. The base size of the anchor boxes are modified compared to the original SSD, as shown in **Table 2**. Those anchor sizes can better fit with our model input size 256x256.

The performance of these ANN object detection networks in detecting and recognizing the road signs are discussed in the Section 3.

3. Results and discussion

3.1 The DIP based algorithms

The DIP based algorithms for detecting and recognizing road signs are implemented in C++ on a Visual Studio platform on desktop computer and a Freescale i.MX6. Due to the lack of standard video datasets dedicated to speed limit signs, we have examined the algorithm using the original video frames captured under different weather conditions such as daylight, backlight, cloudy, night, rain and snow.

3.1.1 System specifications

The DIP based algorithm discussed in Section 2.1 for speed limit and speed regulatory signs are realized on the standard desktop machine consisting of Intel® Core™ i7–3770 CPU, operating at 3.6GHz frequency with a memory of DDR3–1600-8GB on a Windows-7 64 bit and Ubuntu 14.04. The same DIP based algorithms are also realized on the Freescale i.MX 6 which is one of the standard developing processors for the real-time vehicular applications with a ARM Cortex-A9 CPU possessing an operating frequency of 1.2 GHz and memory of 1GB working with a Linux operation system. The same has Video Processing Unit (VPU) decoder-H.264, MPEG-4, H.263, MJPE and Image Processing Unit (IPU) possessing blending, rotating, scaling, cropping, de-interlacing, color spacing converting functions.

3.1.2 Performance: speed

On the desktop computer as well as on the Freescale i.MX 6, the size of images of speed limit signs is set to D1 resolution (720x480). The processing speed of the DIP based algorithm could reach up to 150fps on an average on the desktop computer and about 30fps on the Freescale i.MX 6. The image resolution is set as 1280x720 on the desktop computer and the performance can reach 161 fps on an average. On Freescale i.MX 6, the image resolution is also set as 1280x720 and the performance of the proposed algorithm is 17 fps.

3.1.3 Performance: accuracy and comparison

The performance accuracy of rectangular and circular speed limit signs and triangular speed regulatory signs' detection and recognition by DIP based algorithm discussed in Section 2.1 is tabulated in **Table 3**. The overall accuracy is defined as

	Rectangular speed limit road signs	**Circular speed limit road signs**	**Triangular speed regulatory road signs**
Video Resolution	720x480	720x480	1280x720
Total Video Frames	13187	14332	227445
Total Road Signs Count	77	113	60
Detected Signs	74	108	59
Detection Accuracy	**96.10%**	**95.58%**	**98.33%**
Total Detected Signs Frame Count	429	697	902
Total Detected Signs and Correctly Classified Frames	414	668	853
Total Correctly Recognized Signs Count	72	103	56
Recognition Accuracy	**97.30%**	**96.30%**	**94.92%**
Overall Accuracy (Detection Accuracy * Recognition Accuracy)	**93.51%**	**91.15%**	**93.33%**

Table 3.
The accuracies of the speed limit signs and speed regulatory signs detection and recognition.

"when a car passes a road scene with a road sign instance, the final output of the proposed algorithm is correct as that of the road sign visible to naked eyes."

The detection accuracy of rectangular speed-limit road signs is 96.10% and recognition accuracy is 97.30% accounting to the total accuracy of 93.51%. The detection accuracy of circular speed limit road signs is 95.58% and its recognition accuracy is 96.30% accounting to the overall accuracy of 91.15% whereas the detection accuracy of the triangular speed regulatory signs is 98.33% and the recognition accuracy is 94.92% resulting in the overall accuracy of 93.33%. The performance efficiency of these algorithms are evaluated under different weather conditions such as daytime, cloudy, with strong backlight, nighttime and during snow and rain. Some of these results are tabulated in **Tables 4–6**.

The efficiency of the proposed algorithm is also compared with the state-of-the-art works on the road signs detection and recognition.

As listed in **Table 7**, the proposed speed limit signs detection and recognition system is compared with some of the previous works. It can be implemented on embedded systems for real-time ADAS applications as it is capable of performing under computing resource and support both circular and rectangular speed limit road signs such as 15, 20, 25, 30, 35,....., 110 irrespective of the digit fonts from numerous countries adopting to the blob and breach features.

Video Sequence Number	**1**	**2**	**3**	**4**	**5**	**6**	**7**	**8**
Weather	Day	Day	Day	Day	Cloudy	Rain	Night	Night
Number of Signs	4	4	3	4	1	3	2	2
Detected signs	4	4	3	3	1	3	2	2
Missed Signs	0	0	0	1	0	0	0	0
Number of Frames with Sign Detection	15	17	10	9	4	12	7	9
Number of Correct Speed Limit Sign Recognition	15	16	10	8	3	12	6	9
Number of Wrong Speed Limit Sign Recognition	0	1	0	1	1	0	1	0

Table 4.
Some details of the rectangular speed limit road signs detection and recognition.

Video Sequence Number	**1**	**2**	**3**	**4**	**5**	**6**	**7**	**8**	**9**	**10**
Weather	Day	Day	Day	Day	Backlight	Cloudy	Snow	Rain	Night	Night
Number of Signs	4	6	5	5	5	4	3	2	2	2
Detected signs	4	6	5	5	5	4	2	2	1	2
Missed Signs	0	0	0	0	0	0	1	0	0	0
Number of Frames with Sign Detection	12	20	15	16	16	12	10	10	4	9
Number of Correct Speed Limit Sign Recognition	12	19	15	15	14	11	9	8	4	9
Number of Wrong Speed Limit Sign Recognition	0	1	0	1	2	1	1	2	0	0

Table 5.
Some details of the circular speed limit road signs detection and recognition.

Video Sequence Number	1	2	3	4	5	6	7
Weather	Day	Day	Day	Cloudy	Backlight	Night	Night
Number of Signs	6	3	5	2	4	3	4
Detected signs	5	3	5	2	3	2	4
Missed Signs	1	0	0	0	1	1	0
Number of Frames with Sign Detection	22	13	15	7	14	7	18
Number of Correct Speed Regulatory Sign Recognition	19	12	15	5	14	6	18
Number of Wrong Speed Limit Sign Recognition	3	1	0	2	0	1	0

Table 6.
Some details of the triangular speed regulatory road signs detection and recognition.

	[13]	[14]	[15]	[16]	[23]	[24]	**Speed Limit Road Signs**
CPU	2.16 GHz dual-core laptop	2.13 GHz dual-core laptop	2.13 GHz dual-core laptop	1.167GHz Intel Atom 230 and NVIDIA GeForce 9400 M GS GPU	2.26 GHz dual-core laptop	—	Intel Core i7–3770 3.40 GHz ARM Cortex-A9 1.2 GHz
Video Resolution	700 X 400	Image only	640 X 480	640 X 480	640 X 480	1920 X 1080	720 X 480
Frame Rate on PC	25fps	7.7fps (130 ms)	20fps	33fps	16fps	—	150fps
Detection Accuracy	87.00%	89.68%	97.50%	—	98.90%	—	95.84%
Recognition Accuracy	90.90%	88.97%	96.25%	95.00%	88.50%	96.24%	96.80%
Overall Accuracy	96.25%	90.90%	90.00%	88.00%	98.30%	94.00%	92.10%
Real-Time on Embedded System	X	X	X	X	X	X	O
Supports all types of Speed Limit Signs	X	X	X	X	X	X	O
Supports Different Digit Fonts on Speed Limit Signs	X	X	X	X	X	X	O

Table 7.
The comparison of the proposed speed regulatory road signs detection and recognition algorithm with previous works.

Table 8 lists the comparison of the proposed speed regulatory signs detection and recognition system with relative previous works. It can also be implemented on the embedded systems for real-time ADAS applications, as it is capable of performing under embedded computing resource and support different types of speed regulatory signs as in **Figure 34** from numerous countries adopting feature extraction and feature matching features.

From the comparison listed in the **Tables** 7 and **8**, it can be interpreted that DIP based algorithm discussed in this chapter is more robust i.e., it supports speed limit and speed-regulatory road signs of different types existing in most of the countries. It also performs well with the varied fonts of the speed limit signs and it can be compatibly implemented in an embedded system for the real-time applications. Importantly, it has decent accuracy when working with video as in the real camcorder environment compared to the state-of-the-art methods. Above all, the least complexity of our proposed algorithm yields higher fps compared to those of the other previous works.

Figure 35 shows the experimental results of the speed limit road signs detection and recognition method for detection and recognition of the rectangular speed limit road signs. **Figure 35(a-c)** is the result during the daytime, **Figure 35(d)** is during the cloudy weather, **Figure 35(e-f)** is during the rain and **Figure 35(g-j)** is during the nighttime.

Figure 34.
Some samples of speed regulatory road signs.

	[17]	[18]	[19]	[20]	[21]	**Speed Regulatory Road Signs**
CPU	Intel Core i3	Pentium-IV 2.6 GHz	Intel Core i7	X	Tesla K20 GPU Platform	Intel Core i7–4790 3.60 GHz
Video Resolution	X	640 X 480	1292 X 964	640 X 480	1628 X 1236	1280 X 720
Frame Rate	0.5 fps	11.1 fps	33 fps	13.5 fps	27.9 fps	161 fps
Detection Accuracy	98.25%	97.7%	95.87%	96.00%	91.69%	98.33%
Recognition Accuracy	84.30%	93.60%	99.16%	96.08%	93.77%	94.49%
Overall Accuracy	82.82%	91.44%	95.07%	92.23%	85.97%	93.33%
Sensor Required	Vision	Vision	LIDAR + Vision	Vision	Vision	Vision

Table 8.
The comparison of the proposed speed regulatory road signs detection and recognition algorithm with previous works.

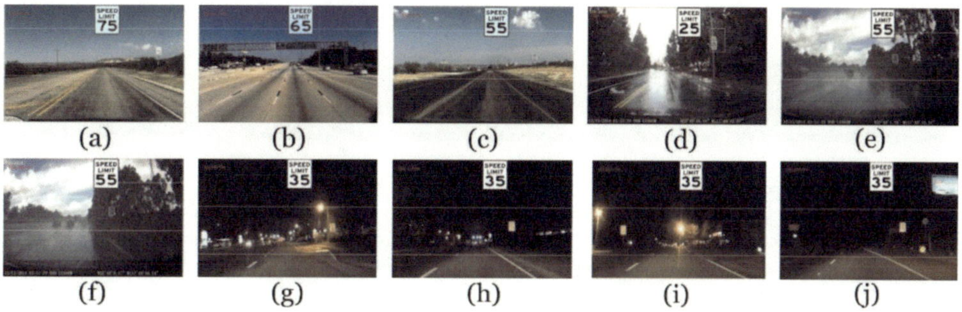

Figure 35.
The overall results for rectangular speed limit signs detection (a-c) during the daytime (d) during the cloudy weather (e-f) during the rain (g-j) during the nighttime.

Figure 36 shows the experimental results of the speed limit road signs detection method for detection as well as recognition of the circular speed limit road signs. **Figure 36(a-c)** is the result during the daytime, **Figure 36(d-e)** is during the backlight condition, **Figure 36(f-g)** is during the cloudy weather, **Figure 36(h-i)** is during the snow, **Figure 36(j-l)** is during the rains and **Figure 36(m-o)** is during the nighttime.

Figure 37 shows the experimental results of the speed regulatory road signs detection and recognition method for detection of the triangular speed regulatory road signs of which **Figure 37(a-c)** is the result during the daytime, **Figure 37(d-g)** is during the cloudy weather, **Figure 37(h-i)** is during the cloudy weather and **Figure 37 (j-l)** is during the nighttime.

Additionally, the proposed CV based method is capable of detecting and recognizing the speed limit signs ending with the digit "5" as in **Figure 38**.

3.2 The CNN based algorithms

The CNN based object detection algorithms such as SSD [51], Faster R-CNN [54], YOLO [55] and the proposed CSPJacinto-SSD are implemented in Python.

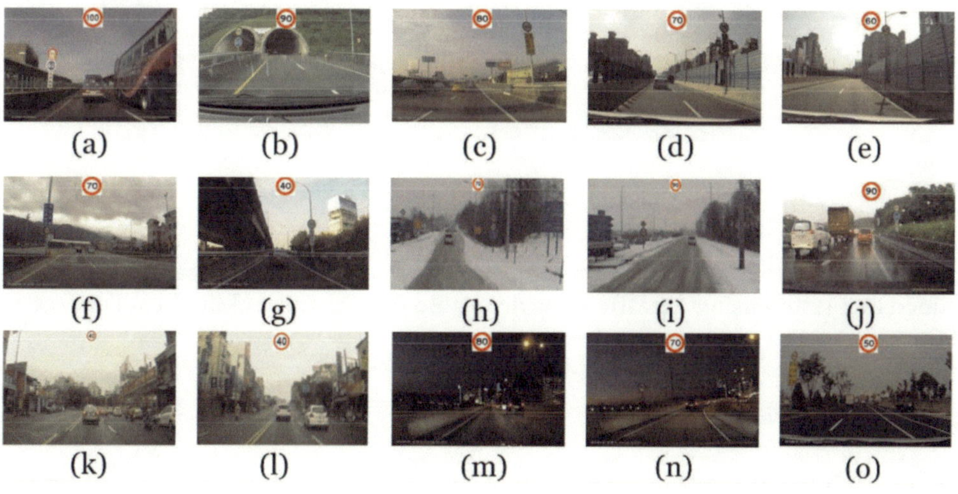

Figure 36.
The overall results for circular speed limit signs detection (a-c) during the daytime, (d-e) during the backlight condition, (f-g) during the cloudy weather, (h-i) during the snow (j-l) during rains, (m-o) during nighttime.

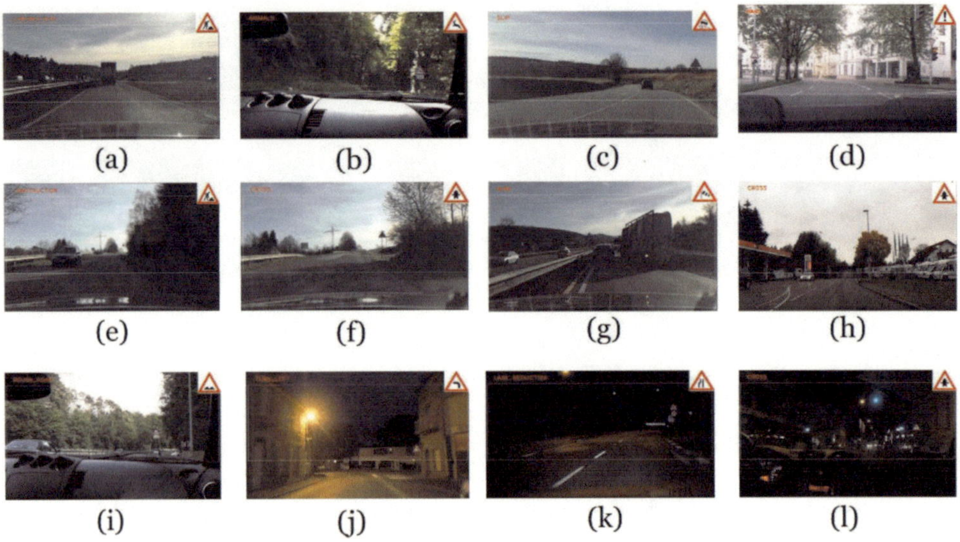

Figure 37.
The overall results for triangular speed regulatory road signs detection (a-c) during the daytime, (d-g) during the backlight condition, (h-i) during the cloudy weather (j-l) during the night.

Figure 38.
The detection and recognition result of speed limit ending with the digit '5'.

In order to carry out the roads signs detection and recognition, the same are trained and tested using a traffic signs dedicated dataset titled 'Tsinghua-Tencent 100 K' [9].

3.2.1 System specifications

The CNN based algorithm discussed in Section 2.1.2 for road signs are realized on the standard desktop machine consisting of Intel® Core™ i7–3770 CPU, operating at 4.2GHz frequency with a memory of DDR3–1600-16GB on a Windows-10 64 bit with Geforce GTX Titan X.

3.2.2 Performance: speed

On the desktop computer, the size of images are used as available in the dataset. The processing speed of the SSD, Faster-RCNN, YOLO and CSPJacinto-SSD object detection algorithms are around 20 fps, 5 fps, 21 fps, and 22 fps respectively.

3.2.3 Performance: accuracy and comparison

The performance efficiency of the CNN models is mostly calculated using mAP, AP and IoU as per [61]. Average precision (AP) is the most commonly used metric to measure the accuracy of object detection by various CNNs, and image-processing methods. The AP computes the average precision value for recall value. Precision measures how accurate the predictions are, by a method, i.e., the percentage of correct predictions whereas, Recall measures the extent to which the predicted positives are good. Eq. (30) is employed in this paper to estimate the AP where r refers to recall rate and $\hat{r}$ refers to the precision value for recall. The interpolated average precision [62] was used to evaluate both classification and detection. The intention in interpolating the precision/recall curve in this way is to reduce the impact of the wiggles in the precision/recall values, caused by small variations in the ranking. Similarly, the mean average precision (mAP) is the average of AP. The accuracy of these models [51, 54, 55] in detecting and recognizing road signs from dataset [9] is as tabulated in **Table 9**.

$$AP = \sum (\mathrm{r_{n+1}} - \mathrm{r_n})\, \mathrm{p_{interp}}\,(\mathrm{r_{n+1}}) \text{ and } \mathrm{p_{interp}}\,(\mathrm{r_{n+1}}) = \max_{\hat{r} \geq rn+1} p(\hat{r}) \tag{30}$$

4. The comparison of DIP and CNN based methods

The traditional DIP based methods are popular CV techniques namely SIFT, SURF, BRIEF, to list a few employed for object detection. Feature extraction process was carried out for image classification jobs. The features are descriptive of "interesting" in images. Various CV algorithms, such as edge detection, corner detection and/or threshold segmentation would be involved in this step. Thus extracted features from images forms the basis for definition for an object to be detected for respective class. During deployment of such algorithms, these definitions are sought in other images. If a significant number of features from defined for a class are found in other images, the image is classified, respectively.

Contrastingly, CNN came up with end-to-end learning where a machine learns about the object from a classes of annotated images which is termed as 'training' of a given dataset. During this, the CNN perceives the fundamental patterns in those

	Input resolution	mAP	FPS	Complexity per frame	No. of parameters
SSD 512	512 x 512	67.90%	~20	105.30G	42.40 M
Faster RCNN	608 x 608	75.20%	~5	120.60G	41.72 M
YOLO V4	608 x 608	71.40%	~21	64.40G	64.50 M
CSPJacinto-SSD	512 x 512	69.60%	~22	18.80G	11.40 M

Table 9.
Performance efficiency of CNN models in detection and recognition of road signs.

classes of images and consistently establishes a descriptive salient features for each specific classes of objects.

With almost all the researches and industries that involves CV are now employing CNN based methods, the functionalities of a CV professional has exceptionally changed in terms of both knowledge, skills and expertise as in **Figure 39**.

A comparison between the DIP and CNN based methods is tabulated in **Table 10**. The DIP based methods are more compatible and suited for the real-time applications of the ADAS as it is of low-complexity and does not require any data for pre-training of the system as compared to the data-hungry neural networks based systems. Apart from pre-training and complexity, the Freescale iMX6 consumes a total power of 1.17 W [64] in the video playback idle mode and other embedded systems would lie in the range where a minimum of 300 W [65] required by a basic GPU which makes them power-hungry as well. Additionally, the DIP based methods can perform object detection in any scenes irrespective of having seen the same or similar scene, but the CNN models can perform object detection of the objects that they models have seen

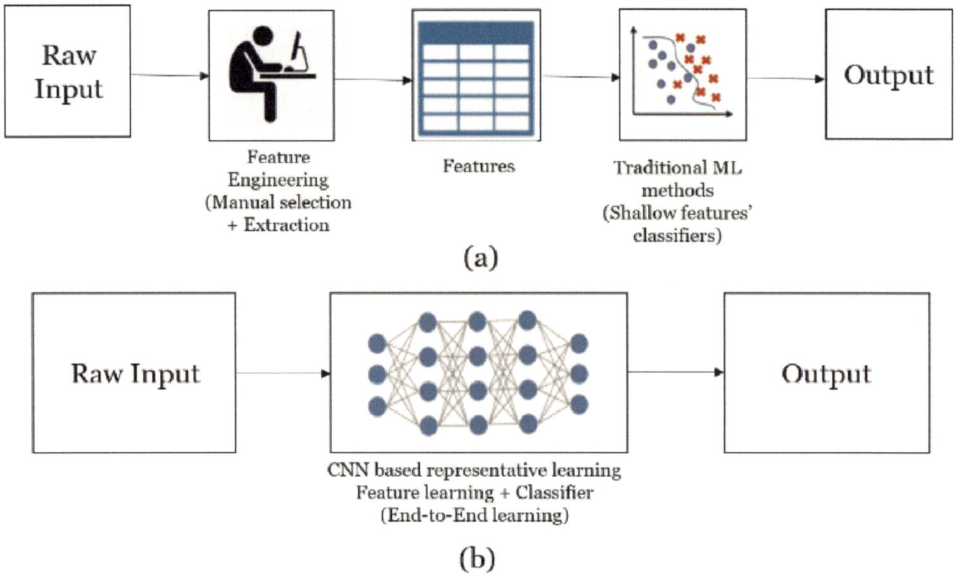

Figure 39.
Comparison of DIP and CNN based workflow. Fig. From [63].

Parameters	Proposed system	CNN based systems
Complexity	Low	High
Pre-training	X	O
Power Consumption	1.17 W	300 W
Recognition in scene never seen before	O	X
Robustness to inclement weathers	X	O

Table 10.
Comparison of the proposed system with that of CNN based systems.

during the training processes. Hence, the CNN models require good-amount of time spent to teach them perform a task followed by testing before employing them for the real-time applications unlike the DIP based methods. Moreover, CNN models exhibit high flexibility and perform better in inclement weathers than DIP based methods.

5. The conclusion

This chapter discussed traditional image processing methods and a few CNN based methods to detection and recognition of road signs for ADAS systems. It has been conclusive that DNNs perform better than the traditional algorithms with certain specific trade-offs with respect to computing requirements and training time. While there are pros and cons of both traditional DIP and CNN based methods, a lot of DIP based CV methods invented over the last 2–3 decades have now become obsolete because newer and much more efficinet methods of CNN have replaced them. However, knowledge and skills gained are also invaluable and not bounded by never inventions instead the knowledge of traditional methods forms a strong foundation for the professional to explore and widen his point-of-viewing a problems. Additionally, there are some of the traditional methods still being used in a hybrid-approach to improvise, innvoate leading to incredible innovations.

Acknowledgements

The authors thank the partial support by the *"Center for mmWave Smart Radar Systems and Technologies"* under the *"Featured Areas Research Center Program"* within the framework of the Higher Education Sprout Project by the Ministry of Education (MOE), Taiwan R.O.C. We also thank the partial support from the Ministry of Science and Technology (MOST), Taiwan R.O.C. projects with grants MOST 108-3017-F-009-001, MOST 110-2221-E-A49-145-MY3, and MOST 109-2634-F-009-017 through Pervasive Artificial Intelligence Research Labs (PAIR Labs) in Taiwan, R.O.C. as well as the partial support from the Qualcomm Technologies under the research collaboration agreement 408929.

Conflict of interest

The authors declare no conflict of interest.

Author details

Vinay M. Shivanna[1*], Kuan-Chou Chen[1], Bo-Xun Wu[1] and Jiun-In Guo[1,2,3]

1 Department of Electronics Engineering and Institute of Electronics, National Yang Ming Chiao Tung University, Hsinchu City, Taiwan

2 Pervasive Artificial Intelligence Research (PAIR) Labs, Hsinchu City, Taiwan

3 Wistron-NCTU Embedded Artificial Intelligence Research Center, Hsinchu City, Taiwan

*Address all correspondence to: vinay.ms23@gmail.com

References

[1] J. Urry, "The 'System' of Automobility", Theory, Culture & Society, vol. 21, no. 4-5, pp. 25-39, October 2004.

[2] E. Eckermann, World History of the Automobile, Society of Automotive Engineers, Warrendale, PA, 2001.

[3] "Transportation: Motor Vehicle Accidents and Fatalities", The 2012 Statistical Abstract. U.S. Census Bureau, September. 2011.

[4] What is Machine Learning[Internet]? Ibm.com. Available from: https://www.ibm.com/cloud/learn/machine-learning

[5] What is Digital Image Processing (DIP) | IGI Global [Internet]. Igi-global.com. Available from: https://www.igi-global.com/dictionary/digital-image-processing-dip/48620

[6] Neural Network Definition[Internet]. Investopedia. Available from: https://www.investopedia.com/terms/n/neuralnetwork.asp

[7] How Artificial Intelligence Works [Internet]. Investopedia. Available from: https://www.investopedia.com/terms/a/artificial-intelligence-ai.asp

[8] Houben S, Stallkamp J, Salmen J, Schlipsing M, Igel C. Detection of traffic signs in real-world images: The German traffic sign detection benchmark. The 2013 International Joint Conference on Neural Networks (IJCNN). IEEE; 2013. p. 1-8. Available from: http://10.1109/IJCNN.2013.6706807

[9] Zhu Z, Liang D, Zhang S, Huang X, Li B, Hu S. Traffic-Sign Detection and Classification in the Wild. 2016 IEEE Conference on Computer Vision and Pattern Recognition (CVPR) . IEEE; 2016. p. 2110-2118. Available from: http://10.1109/CVPR.2016.232

[10] Deng J, Dong W, Socher R, Li L, Li K, Li F. ImageNet: A large-scale hierarchical image database. 2009 IEEE Conference on Computer Vision and Pattern Recognition. IEEE; 2009. p. 248–255. Available from: http://10.1109/CVPR.2009.5206848.

[11] Everingham M, Van Gool L, Williams C, Winn J, Zisserman A. The Pascal Visual Object Classes (VOC) Challenge. International Journal of Computer Vision. 2009;88(2): 303-338.

[12] Mechanical Simulation [Internet]. Carsim.com. Available from: https://www.carsim.com/

[13] Jim Torresen, Jorgen W. Bakke and Lukas Sekanina, "Efficient Recognition of Speed Limit Signs," Proc. 2004 IEEE Intelligent Transportation Systems Conference, Washington, D.C., USA, October 3-6, 2004.

[14] Fabien Moutarde, Alexandre Bargeton, Anne Herbin, and Lowik Chanussot, "Robust on-vehicle real-time visual detection of American and European speed limit signs, with a modular Traffic Signs Recognition system," Proc. 2007 IEEE Intelligent Vehicles Symposium, Istanbul, Turkey, June 13-15, 2007.

[15] Christoph Gustav Keller, Christoph Sprunk, Claus Bahlmann, Jan Giebel and Gregory Baratoff, "Real-time Recognition of U.S. Speed Signs," Proc. 2008 IEEE Intelligent Vehicles Symposium, June 4-6, 2008, The Netherlands.

[16] Wei Liu,Jin Lv, Haihua Gao, Bobo Duan, Huai Yuan and Hong Zhao, “An Efficient Real-time Speed Limit Signs Recognition Based on Rotation Invariant Feature”, Proc. 2011 IEEE Intelligent Vehicles Symposium (IV) Baden-Baden, Germany, June 5-9, 2011.

[17] Zumra Malik and Imran Siddiqi, “Detection and Recognition of Traffic Sign Road Scene Images”, 12th International Conference on Frontiers of Information Technology, pp 330-335, Dec 17-19, 2014.

[18] Vavilin Andrey and Kang Hyun Jo, “Automatic Detection and Recognition of Traffic Signs using Geometric Structure Analysis”, International Joint Conference on SICE-ICASE, Oct 18-21, 2006.

[19] Lipu Zhou, Zhidong Deng, “LIDAR and Vision-Based Real-Time Traffic Sign Detection and Recognition Algorithm for Intelligent Vehicle”, IEEE 17th International Conference on Intelligent Transportation Systems (ITSC), Oct 8-11, 2014.G. Eason, B. Noble, and I. N. Sneddon, “On certain integrals of Lipschitz-Hankel type involving products of Bessel functions,” Phil. Trans. Roy. Soc. London, vol. A247, pp. 529–551, April 1955. *(references)*

[20] Sebastian Houben, Johannes Stallkamp, Jan Salmen, Marc Schlipsing, and Christian Igel, “Detection of Traffic Signs in Real-World Images: The German Traffic Sign Detection Benchmark” The International Joint Conference on Neural Networks (IJCNN), 2013.

[21] M. liang, M. Yuan, X. Hu, J. Li, and H. Liu, “Traffic sign detection by supervised learning of color and shape,” in Proceedings of IEEE International Joint Conference on Neural Networks, 2013.

[22] M. Mathias, R. Timofte, R. Benenson, and L. V. Gool, “Traffic sign recognition - how far are we from the solution?” in Proceedings of IEEE International Joint Conference on Neural Networks, 2013.

[23] G. Wang, G. Ren, Z. Wu, Y. Zhao, and L. Jiang, “A robust, coarse-to-fine traffic sign detection method,” in Proceedings of IEEE International Joint Conference on Neural Networks, 2013.

[24] Supreeth H.S.G, Chandrashekar M Patil, “An Approach Towards Efficient Detection and Recognition of Traffic Signs in Videos using Neural Networks” International Conference on Wireless Communications, Signal Processing and Networking (WiSPNET), 2016, pp 456-459

[25] Nadra Ben Romdhane, Hazar Mliki, Mohamed Hammami, “An mproved Traffic Signs Recognition and Tracking Method for Driver Assistance System”, IEEE/ACIS 15th International Conference on Computer and Information Science (ICIS)-2016

[26] B. Lucas and T. Kanade, “An iterative image registration technique with an application to stereo vision”, The International Joint Conference on Artificial Intelligent, 1981, pp. 674–679.

[27] Y. Zhang, C. Hong and W. Charles, "An efficient real time rectangle speed limit sign recognition system," 2010 IEEE Intelligent Vehicles Symposium, San Diego, CA, 2010, pp. 34-38. DOI: 10.1109/IVS.2010.5548140

[28] A. Mammeri, A. Boukerche, J. Feng and R. Wang, "North-American speed limit sign detection and recognition for smart cars," 38th Annual IEEE Conference on Local Computer Networks - Workshops, Sydney, NSW, 2013, pp. 154-161

[29] C. Wang, "Research and Application of Traffic Sign Detection and Recognition Based on Deep Learning," 2018 International Conference on Robots & Intelligent System (ICRIS), 2018, pp. 150-152, doi: 10.1109/ ICRIS.2018.00047.

[30] R. Hasegawa, Y. Iwamoto and Y. Chen, "Robust Detection and Recognition of Japanese Traffic Sign in the Complex Scenes Based on Deep Learning," 2019 IEEE 8th Global Conference on Consumer Electronics (GCCE), 2019, pp. 575-578, doi: 10.1109/ GCCE46687.2019.9015419.

[31] Y. Sun, P. Ge and D. Liu, "Traffic Sign Detection and Recognition Based on Convolutional Neural Network," 2019 Chinese Automation Congress (CAC), 2019, pp. 2851-2854, doi: 10.1109/ CAC48633.2019.8997240.

[32] Y. Yang, H. Luo, H. Xu and F. Wu, "Towards Real-Time Traffic Sign Detection and Classification," in IEEE Transactions on Intelligent Transportation Systems, vol. 17, no. 7, pp. 2022-2031, July 2016, doi: 10.1109/ TITS.2015.2482461.

[33] R. Jain and D. Gianchandani, "A Hybrid Approach for Detection and Recognition of Traffic Text Sign using MSER and OCR," 2018 2nd International Conference on I-SMAC, 2018, pp. 775-778, doi: 10.1109/I-SMAC.2018.8653761.

[34] M. Z. Abedin, P. Dhar and K. Deb, "Traffic sign recognition using hybrid features descriptor and artificial neural network classifier," 2016 19th International Conference on Computer and Information Technology (ICCIT), 2016, pp. 457-462, doi: 10.1109/ ICCITECHN.2016.7860241.

[35] Lin Y, Chou T, Vinay M, Guo J. Algorithm derivation and its embedded system realization of speed limit detection for multiple countries. 2016 IEEE International Symposium on Circuits and Systems (ISCAS). Montreal, QC: IEEE; 2016. p. 2555-2558. Available from: http://10.1109/ ISCAS.2016.7539114

[36] Chou T, Chang S, Vinay M, Guo J. Triangular Road Signs Detection and Recognition Algorithm and its Embedded System Implementation. The 21st Int'l Conference on Image Processing, Computer Vision and Pattern Recognition. CSREA Press; 2017. p. 71-76. Available from: http://ISBN: 1-60132-464-2

[37] Gareth Loy and Nick Bames, "Fast Shape-based Road Sign Detection for a Driver Assistance System," Proc. IEEE/ RSl International Conference on Intelligent Robots and Systems, September 28 - October 2, 2004.

[38] Nick Barnes and Gareth Loy, "Real-time regular polygonal sign detection", Springer Tracts in Advanced Robotics Volume 25, pp 55-66, 2006.

[39] Sebastian Houben, "A single target voting scheme for traffic sign detection," Proc. 2011 IEEE Intelligent Vehicles Symposium (IV), Baden-Baden, Germany, June 5-9, 2011.

[40] Feature Detectors - Sobel Edge Detector. Homepages.inf.ed.ac.uk. Available from: https://homepages.inf.ed .ac.uk/rbf/HIPR2/sobel.htm

[41] Fatin Zaklouta and Bogdan Stanciulescu, "Real-time traffic sign recognition in three stages," Robotics and Autonomous Systems, Volume 62, Issue 1, January 2014.

[42] Derek Bradley and Gerhard Roth, "Adaptive Thresholding using the Integral Image," Journal of Graphics,

GPU, and Game Tools, Volume 12, Issue 2, 2007.

[43] Alexandre Bargeton, Fabien Moutarde, Fawzi Nashashibi, and Benazouz Bradai, "Improving pan-European speed-limit signs recognition with a new "global number segmentation" before digit recognition," Proc. 2008 IEEE Intelligent Vehicles Symposium, Eindhoven, The Netherlands, June 4-6, 2008.

[44] Dang Khanh Hoa, Le Dung, and Nguyen Tien Dzung, "Efficient determination of disparity map from stereo images with modified Sum of Absolute Differences (SAD) algorithm", 2013 International Conference on Advanced Technologies for Communications (ATC 2013)

[45] J. R. Parker, "Vector Templates and Handprinted Digit Recognition," Proc. 12th IAPR International Conference on Pattern Recognition, 1994. Vol. 2 - Conference B: Computer Vision & Image Processing, Jerusalem, 9-13 Oct 1994.

[46] Phalgun Pandya and Mandeep Singh, "Morphology Based Approach To Recognize Number Plates in India," International Journal of Soft Computing and Engineering (IJSCE), Volume-1, Issue-3, July 2011.

[47] Kamaljit Kaur and Balpreet Kaur, "Character Recognition of High Security Number Plates Using Morphological Operator," International Journal of Computer Science & Engineering Technology (IJCSET), 2011 IEEE Intelligent Vehicles Symposium (IV), Vol. 4, May, 2013.

[48] Lifeng He and Yuyan Chao, "A Very Fast Algorithm for Simultaneously Performing Connected-Component Labeling and Euler Number Computing", IEEE TRANSACTIONS ON IMAGE PROCESSING, VOL. 24, NO. 9, SEPTEMBER 2015

[49] Rachid Belaroussi and Jean Philippe Tarel, "Angle Vertex and Bisector Geometric Model for Triangular Road Sign Detection", IEEE Winter Conference on Applications of Computer Vision (WACV), pp 1-7, 2009.

[50] H. Bay, T. Tuytelaars, and L. Van Gool, "Surf: Speeded up robust features", European Conference on Computer Vision, May 2006.

[51] Liu W, Anguelov D, Erhan D, Szegedy C, Reed S, Fu C et al. SSD: Single Shot MultiBox Detector. In Proceedings of the European Conference on Computer Vision (ECCV). 2016. p. 21–37. Available from: http://arXiv: 1512.02325

[52] R. Girshick, J. Donahue, T. Darrell, and J. Malik. Rich feature hierarchies for accurate object detection and semantic segmentation. In CVPR, 2014

[53] Girshick R. Fast R-CNN. 2015 IEEE International Conference on Computer Vision (ICCV) [Internet]. IEEE; 2015. Available from: http://10.1109/ ICCV.2015.169

[54] I. Goodfellow, J. Pouget-Abadie, M. Mirza, B. Xu, D. Warde-Farley, S. Ozair, A. Courville, and Y. Bengio, "Generative adversarial nets," in Advances in Neural Information Processing Systems, 2014, pp. 2672– 2680.

[55] Ren S, He K, Girshick R, Sun J. Faster R-CNN: Towards Real-Time Object Detection with Region Proposal Networks. IEEE Transactions on Pattern Analysis and Machine Intelligence. 2017; 39(6):1137-1149.

[56] I. Goodfellow, J. Pouget-Abadie, M. Mirza, B. Xu, D. Warde-Farley, S. Ozair,

A. Courville, and Y. Bengio, "Generative adversarial nets," in Advances in Neural Information Processing Systems, 2014, pp. 2672– 2680.

[57] Martin Arjovsky, Soumith Chintala, and LeonBottou.Wasserstein GAN. arXiv:1701.07875v2 [stat.ML], 9 Mar 2017.

[58] C. Wang, H. M. Liao, Y. Wu, *P. Chen*, J. Hsieh, and I. Yeh, "CSPNet: A New Backbone that can Enhance Learning Capability of CNN," in 2020 IEEE/CVF Conference on Computer Vision and Pattern Recognition Workshops (CVPRW), 14-19 June 2020 2020, pp. 1571-1580.

[59] M. Mathew, K. Desappan, P. K. Swami, and S. Nagori, "Sparse, Quantized, Full Frame CNN for Low Power Embedded Devices," in 2017 IEEE Conference on Computer Vision and Pattern Recognition Workshops (CVPRW), 21-26 July 2017 2017, pp. 328-336

[60] C. Y. Lai, B. X. Wu, T. H. Lee, V. M. Shivanna, and J. I. Guo, "A Light Weight Multi-Head SSD Model For ADAS Applications," in 2020 International Conference on Pervasive Artificial Intelligence (ICPAI), 3-5 Dec. 2020 2020, pp. 1-6

[61] Jonathan, H. mAP (mean Average Precision) for Object Detection. Available online: **https://medium.com/@jonathan_hui/map-mean-average-precision-for-object-detection-45c121a31173** (accessed on 12 July 2018).

[62] Salton, G.; McGill, M.J. *Introduction to Modern Information Retrieval*; McGraw-Hill: New York, NY, USA, 1986. [**Google Scholar**]

[63] Wang J, Ma Y, Zhang L, Gao RX (2018) Deep learning for smart manufacturing: Methods and applications. J Manuf Syst 48:144–156. https://doi.org/10.1016/J.JMSY.2018.01.003

[64] Freescale Semiconductor, "i.MX 6Dual/6Quad Power Consumption Measurement" from "https://bit.ly/2ATVcWk"

[65] GEFORCE. "Desktop GPUs-Specifications" from "https://www.geforce.co.uk/hardware/desktop-gpus/geforce-gt-1030/specifications"

Chapter 7

Smart-Road: Road Damage Estimation Using a Mobile Device

Izyalith E. Álvarez-Cisneros, Blanca E. Carvajal-Gámez, David Araujo-Díaz, Miguel A. Castillo-Martínez and L. Méndez-Segundo

Abstract

Mexico is located on five tectonic plates, which when moving, generate telluric movements. These movements, depending on their intensity, affect the telecommunications infrastructure. Earthquakes tend to cause landslides, subsidence, damage to structures in houses, buildings, and roads. In the case of road damage, it is reflected in cracks in the pavement, which are classified according to their size, shape, and depth. The methods that are currently implemented to inspect roads mainly use human perception and are limited to a superficial inspection of the terrain, causing this process ineffective for the timely detection of damage. This work presents a method of road analysis using a drone to acquire images. For the processing and recognition of damages, a mobile device is used, allowing to determine the damage type on the road. Artificial intelligence techniques are implemented to classify them into linear cracks or zig-zag cracks.

Keywords: convolutional neural networks, computational vision, descriptors, cracks road, earthquakes

1. Introduction

A country that is endowed with good road infrastructures can generate the basic elements of competitiveness and provide opportunities for better economic development and at the same time to promote its social and cultural development [1]. There are several factors by which access roads can be affected, and some examples are as follows: time, use, excessive weight, the quality of materials, ubication, natural disasters, etc. We can highlight the damage to the roads caused by earthquakes, which is due to the movement of the tectonic plates, it can cause fissures that are on the surface. Within the damages caused on the roads, deterioration can occur from small cracks too wide ruptures or separations in the road, and these types of incidents tend to occur mainly in the seismic areas of the country.

1.1 Seismicity in Mexico

The Mexican Republic is in located one of the most seismically active regions in the world and is immersed within the area known as the Circumpacific Belt (or Pacific

Ring of Fire), where the greatest seismic and volcanic activity on the planet is concentrated [2], **Figure 1**.

The Global Seismic Hazard Assessment Program (GSHAP) was a project sponsored by the United Nations (UN) that assembled the first worldwide map of earthquake zones [3]. In Mexico, an earthquake hotspot follows the route through the Sierra Madre Occidental that reaches south of Puerto Vallarta to the Pacific coast on the border with Guatemala [3]. **Figure 2** shows the seismic regionalization in the Mexican Republic, marking the zone A as the one with the lowest risk, followed by zones B, C, and D. These last three are the ones that generate the greatest concern to the scientific community and the inhabitants of these areas, due to the structural damage that occurs on the roads and in their localities.

Figure 1.
Circumpacific Belt Zone. Source: National Institute of Statistic and Geography.

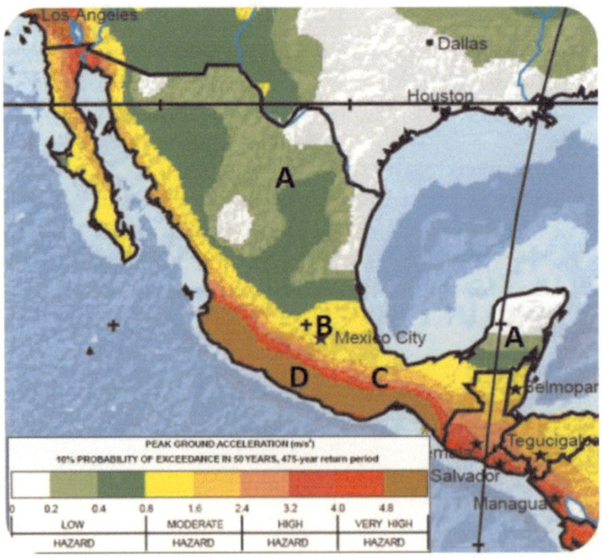

Figure 2.
Areas with the highest propensity to earthquakes in Mexico [3].

1.2 Road network in Mexico

In Mexico, as in other countries, the road network is the most widely used transport infrastructure. The national network has 378,923 km, which is made up of avenues, streets, highways, and rural roads that allow connectivity between practically all the populations of the country. **Figure 3** shows the main roads that interconnect the Mexican Republic [4].

1.2.1 Damage to land access roads

Seismic activity is recurrent in certain areas, causing damage to road infrastructure. These damages are identified as fissures, cracks in the asphalt, landslides, separation of road sections, subsidence, and other damages in different access roads that interconnect the country. The road must be inspected, and the damages detected must be reported and repaired. **Figure 4** shows some examples of damage caused by seismicity in Mexican territory on different roads [5].

Figure 3.
Mexico's major highways 2009 [4].

Figure 4.
Damage to roads caused by earthquakes, left: Chiapas, 5.6°r earthquake, right: Oaxaca, 5.2°r. Source: Google.

In Mexico, most of the road inspections after an earthquake are carried out in person, which can generate more conflicts in some critical points. In these conflict points, semi-autonomous surveillance systems are required that implement mobile technology to detect damage to access roads. Therefore, it is proposed to develop a methodology that, through image processing techniques and neural networks, allows the identification of damage to roads, classifying two types of cracks: linear and zig-zag. This chapter is divided into six sections. Section 2 explains the related work, Section 3 explain methods and materials, Section 4 provides tests and results, Section 5 conclusion and finally Section 6 discussion.

2. Related work

There is research in the field of artificial intelligence related to techniques and practices used to automate the detection of road defects. Below are some related works that have been developed: In [6], an automatic system for identifying cracks in roads through a camera is developed. It scans the roads by zone and inspects the condition of cracks and fissures. The authors propose the following stages: i) smooth, adjust, and binarize the image using the threshold value method, ii) perform morphological operations such as dilation and erosion, iii) eliminate false cracks in the image with smoothing filters, iv) clean and perform the connection of cracks in the image, and finally v) estimate the shape of the crack, using geometric characteristics and shape description. In [7], using texture classifiers, the authors address the descriptors. Through these techniques, it is possible to detect color and texture changes in an image, and thus perform the identification of edges by extracting a set of characteristics, generated from these histograms. For each frame of a pavement video analyzed, the method extracts the characteristic and creates its binary version to classify each region.

In [8], the authors apply morphological operations to the images and segment the images using the K near-neighbor (Knn) method. The proposed algorithm highlights the information of the image texture, and the results are classified using the standard deviation; to define regions delimited by the intensity of gray, these techniques allow to detect patches on the roads through images processed on a Smartphone. **Figure 5** shows the results presented by the authors.

In [9], a system for identifying cracks in buildings from an unmanned aerial vehicle (UAV) equipped with a camera is presented. Fly through the building to acquire images, which are transmitted remotely *via* Wi-Fi to a computer for processing. Images are segmented with techniques to change the Red, Green, Blue (RGB) color space to grayscale. The threshold is calculated with statistical methods (mean and standard deviation), to categorize the black and white pixels and identify the cracks in the building.

Maeda et al. [9] developed a system for identifying cracks in the pavement, where images are captured from a Smartphone mounted on a cell phone holder on the dashboard of a car. It develops an application that analyzes images obtained from the Smartphone through a deep neural network that allows the identification of cracks in the road. In this work, they use deep neural networks such as Region-based Convolutional Neural Networks (R-CNN), You Only Look Ones (YOLO), and Single Shot MultiBox Detector (SSD), for the extraction of characteristics from the region of interest (cracks). In 2019, Zhang et al. [10] propose an intelligent monitoring system to evaluate the damage in the pavement, and this methodology proposes the use of a

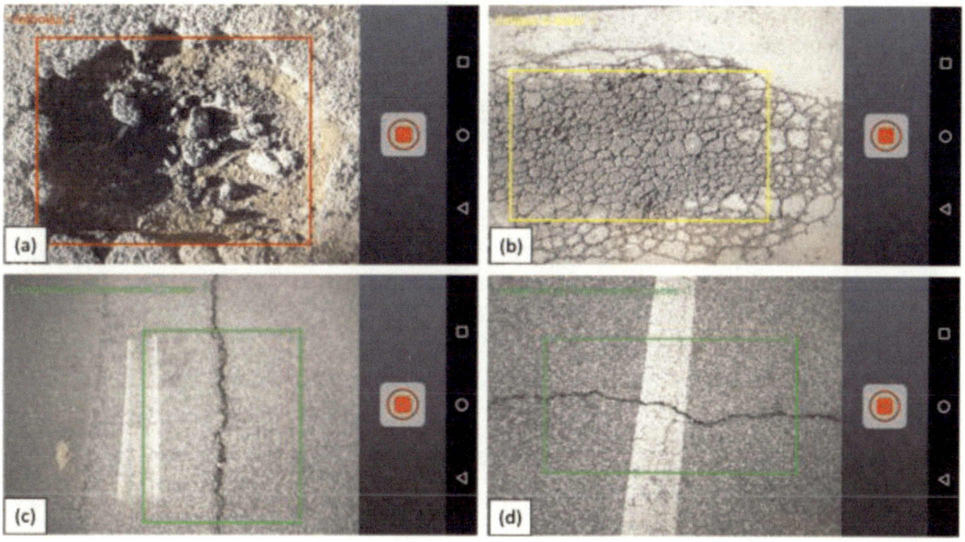

Figure 5.
Detection of the mobile damage system. a) Hole in the pavement, b) longitudinal crack, c) transverse crack, and d) horizontal crack [8].

set of points of an image obtained from a UAV, making use of Harris performing the processing in the cloud for the identification of cracks in the pavement.

3. Methods and materials

In **Figure 6**, the general architecture of the proposed methodology for the classification and identification of linear and zig-zag cracks is shown.

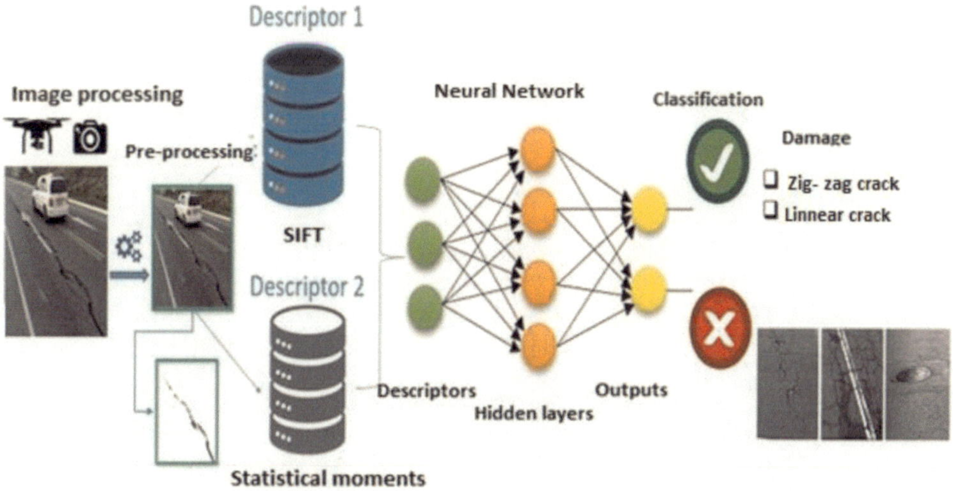

Figure 6.
Proposed architecture for the identification and classification of cracks.

Figure 6 shows the methodology is composed of different stages of development which are image acquisition, pre-processing, descriptors, classification, and result. Each of these stages is detailed below.

3.1 Image acquisition

In this step, the image is taken with the camera that the PARROT BEBOP 2 FPV Drone has, which has the following characteristics: 14-megapixel camera with wide-angle lens, unique digital image stabilization system, live video from a Smartphone or tablet with a viewing angle of 180°, photo format: RAW, JPEG, DNG, and image resolution of 3800 × 3188 pixels to automate the route of the drone, a function implemented to trace the flight path is used. In **Figure 7**, the programmed route map is shown.

3.2 Pre-processing

3.2.1 Dimension reduction

In this section, the image scaling is performed by implementing the Discrete Wavelet Transform "Haar" (DWT-H). In **Figure 8**, three levels of decomposition are shown.

3.2.2 Edge enhancement

To obtain the edge enhancement of the image obtained from point 3.2.1, the use of a Laplacian filter is proposed. The Laplacian of an image highlights the regions of rapid intensity change and is an example of a second-order or a second derivative method of enhancement. It is particularly good at finding the fine details of an image. Any feature with a sharp discontinuity will be enhanced by a Laplacian operator [11]. The Laplacian is a well-known linear differential operator approximating the second derivative given by Eq. (1).

Figure 7.
Simulation of a programmed route at 15 mts.

Figure 8.
DWT-H decomposition.

$$\nabla^2 f = \frac{\partial^2 f}{\partial x^2} + \frac{\partial^2 f}{\partial y^2} \tag{1}$$

where f denotes the image. The following process is performed, a 3×3 matrix is convolved with the image, **Figure 9**.

In **Figure 10**, the Laplacian filtering process is shown. This process consists of the following steps: From the image obtained by the DWT-H (**Figure 10a**), the convolution is performed with the proposed 3×3 kernel (**Figure 10b**). Finally, the sub-image of the crack is obtained with the edges highlighted as seen in **Figure 10d**.

3.3 Feature extraction

One of the main objectives of this work is to implement the methodology on a mobile device, which will perform the image processing offline, obtaining the result

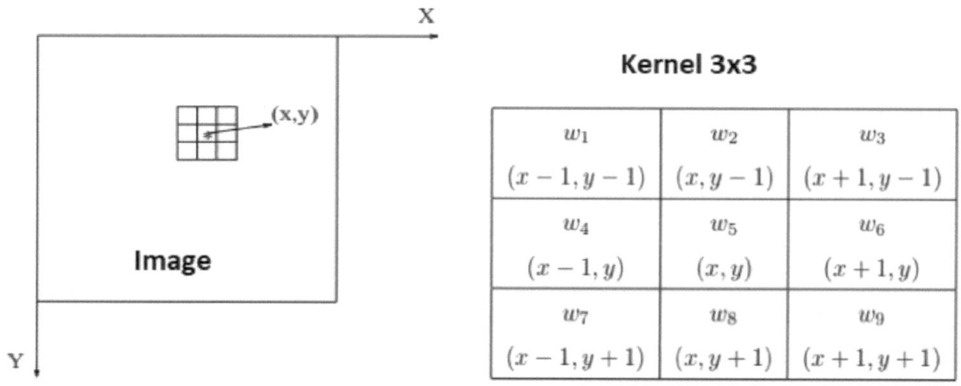

Figure 9.
Convolution of the 3×3 kernel at a point (x, y) in the image.

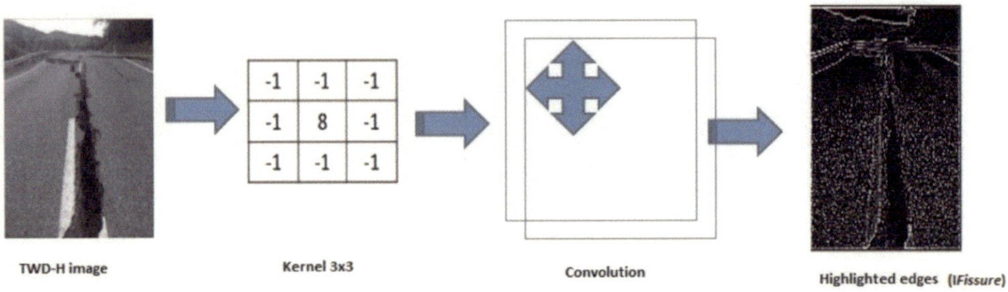

Figure 10.
The result obtained by the Laplacian Filter.

on the site. It is therefore essential to extract only the key points that provide information about features outstanding image and thus make their classification in a convolutional neural network LeNet efficient. To do this, it is proposed to perform the extraction of the characteristics through the scale-invariant feature transform (SIFT) and the pixel rearrangement of the points thrown from the Laplacian filter through statistical moments. The following is the extraction of the characteristics:

3.3.1 Statistical central moments

Central moments also referred to as moments of the mean has been calculated as [12], Eq. (2),

$$\mu_m = \sum_{n=0}^{L-1}((X_n - y)^m t(X_n)) \tag{2}$$

where 'm' is the order of the moment, 'L' is the number of possible intensity values, 'X_n' is the discrete variable that represents the intensity level in the image, and 'y' is the mean of the values, $t(X_n)$ is the probability estimate of the occurrence of 'X_n', Eq. (3).

$$y = \sum_{n=0}^{L-1}(X_n t(X_n)) \tag{3}$$

The mean is the first-order moment followed by variance, skewness, and kurtosis as the second, third, and fourth moments. The mean at the first-order central moment is used to measure the average intensity value of the pixel distribution. Variance (μ2) was used to measure how wide the pixels spread over from the mean value, Eq. (4).

$$\mu_2 = \sum_{n=0}^{L-1}\left((X_n - y)^2 t(X_n)\right) \tag{4}$$

To know the dispersion of the values located as key points by SIFT, the second central moment is implemented to group the pixels of the image processed by the Laplacian Filter. The smoothness texture "R" is defined by Eq. (5),

$$R = \left(1 - \left(1/_{(1+\mu_2(x))}\right)\right) \tag{5}$$

where 'μ2' is the variance and '*x*' is an intensity level. Then, the following condition is established, by Eq. (6),

$$I_{Fissure}(i,j) = \begin{cases} R - y = 1, & I_{Fissure} = 1 \\ OTHERWISE, & I_{Fissure} = 0 \end{cases} \quad (6)$$

3.3.2 SIFT

According to the SIFT methodology [13], the first step is scale detection. For the particular case of the crack contour, this step is very useful for the identification of the crack, since the taking of the images can vary depending on the shooting distance. The formal description of this step is detailed below.

3.3.2.1 Scale detection

The scalar space L (x, y) of an image, is obtained from the convolution of an input image $I_{Fissure}$, through a Gaussian filter G (x, y, σ) at different scales of the value of σ = 0.5 [13], Eq. (7):

$$L(x,y,\sigma) = G(x,y,\sigma) * I_{Fissure}(x,y,\sigma) \quad (7)$$

where, $(x,y,\sigma) = \frac{1}{2\pi\sigma^2} e^{\frac{-(x^2+y^2)}{2\sigma^2}}$, is the function of the Gaussian filter; it is applied in both dimensions (x,y) of the $I_{Fissure}$ image plane.

To obtain the different scale versions of the $I_{Fissure}$ image, it is necessary to multiply the value of σ with different values of the constant k to obtain the projections of the contiguous scales (where k takes values k > 1), each scale's projection is subtracted with the original scale, obtaining the differences from the original image $I_{Fissure}$, Eq. (8):

$$D_m(x,y,\sigma) = L(x,y,k\sigma) - L(x,y,\sigma) \quad (8)$$

The search for extreme values on the spatial scale produces multiple candidates of which the points that are not considered are the low contrast ones since they are not stable to changes in lighting and noise. Eq. (9) shows how the points of interest are located within the image and these locations are given by [13]:

$$z = -\frac{\partial^2 D^{-1}(x,y,\sigma)}{\partial x^2} \frac{\partial D(x,y,\sigma)}{\partial x} \quad (9)$$

Subsequently, the vectors are arranged according to the orientation of the points obtained from Eq. (9), and it is explained below.

3.3.2.2 Orientation mapping

This step assigns a constant orientation to the key points based on the properties of the image obtained in the previous steps. The key point descriptor can be represented with this orientation, achieving the invariance to rotation, which is important to highlight because the image can be taken at different shooting angles. The procedure to find the orientation of the points, is as follows [13]:

Using the scalar value of the points of interest selected in Eq. (4).

a. Calculation of the magnitude value, M.

$$M(x,y) = \sqrt{\left((L(x+1,y) - L(x-1,y))^2 + (L(x,y+1) - L(x,y-1))^2\right)} \quad (10)$$

b. Calculation of orientation, θ

$$\theta(x,y) = tan^{-1}((L(x,y+1) - L(x,y-1)/(L(x+1,y) - L(x-1,y)))) \quad (11)$$

Finally, the description of the characteristic points obtained in the previous steps must be identified the interesting points, **Figure 11**.

3.3.3 Convolutional neural network LeNet

The neural network will allow, based on the characteristics obtained, to train and identify the cracks that appear in the image. The neural network used in this research is the convolutional neural network LeNet, which is made up of five layers of neurons in its architecture, and has an input of 1024 × 1024 × 3 values, and an output of two possible classes [14]. LeNet is a network that is optimized for mobile devices, which allows greater efficiency in the detection and performance of the processes on the mobile device. The network architecture is presented in below **Figure 12** [14]:

For the training of the network, a collection of approximately 500 images were made in various areas of Tecámac, State of Mexico. Therefore, in this investigation,

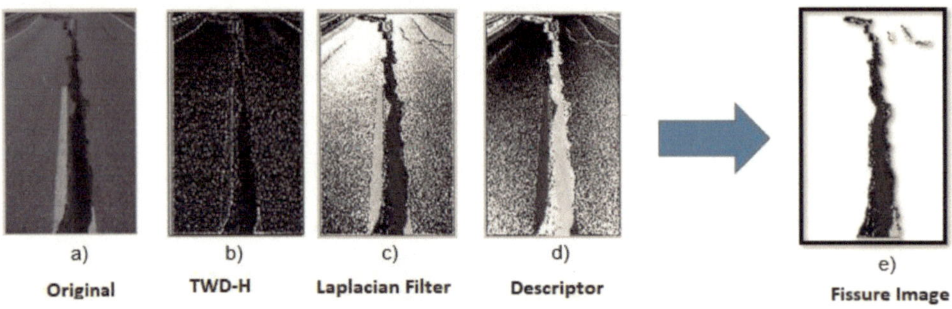

Figure 11.
Results obtained from the proposed methodology: a) original image, b) image obtained with the DWT-H, c) image with the Laplacian filter, d) image obtained with the descriptors, and e) final image.

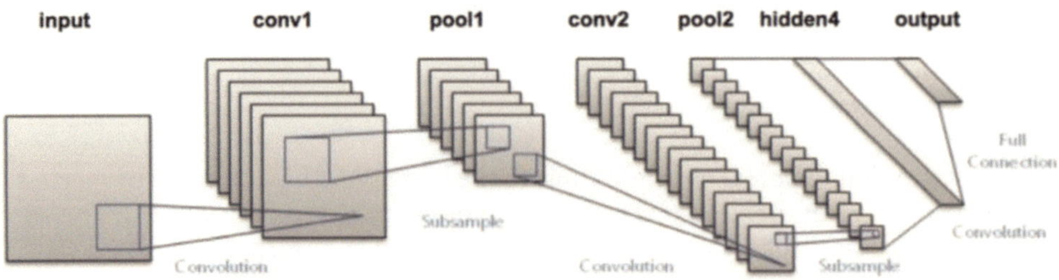

Figure 12.
LeNet neural network architecture [14].

Figure 13.
Zig-zag cracks.

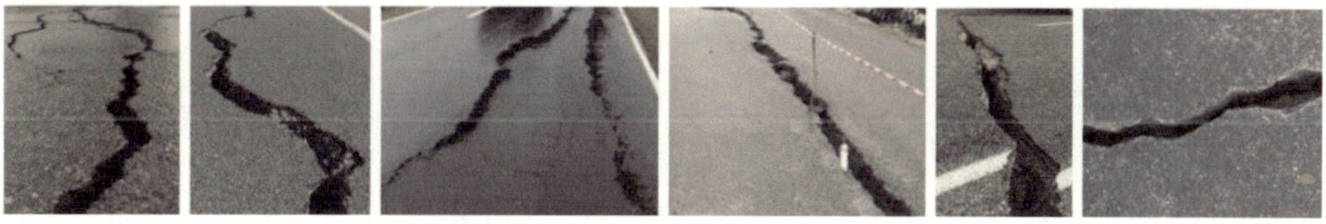

Figure 14.
Linear cracks.

cracks with different intensities will be detected, so these will be identified and classified in the following categories [15]:

Erratic or zigzag cracks (ZZC): These types of cracks in the pavement with erratic longitudinal patterns. It is presented by extreme changes in temperature, defective base, and seismic movements.

Significant cracks (LC): These are cracks with a length greater than 30 centimeters.

Very significant cracks (VSC): Those that are shown in the pavement and have a length greater than 60 centimeters due to their size are a risk. These cracks are the most visible.

Non-significant cracks (NSC): These are cracks that appear in the pavement and that have a fine shape and a length of fewer than 30 centimeters. **Figures 13** and **14** show images referring to the classifications that have been delimited for identification. These define the two classes to detector zig-zag crack (ZZC) and linear crack (LC), respectively.

4. Tests and results

To perform the tests, they were divided into phases to estimate the time of each one of these and thus detect which of them generates a greater consumption of time compared to the others. The processing and results were developed on a Motorola X4 mobile device with a processor: 2.2 GHz and 3 GB RAM. To select the optimum distance for taking the images, tests were carried out between 10 and 30 meters above ground level. At each distance, it was ensured that the images were clear and that the crack would be visualized. **Figure 15** shows the range in height and image visibility for the 500 sample images. From **Figure 15**, we can see that at a height of 10 meters the drone has a visibility range of 26 meters in radius. In a similar way we can observe that for heights of 15, 20, 25 and 30 meters, they correspond to 40, 53, 67 and 80 meters of visibility radius. For our case we consider a height between 15 and 20 meters.

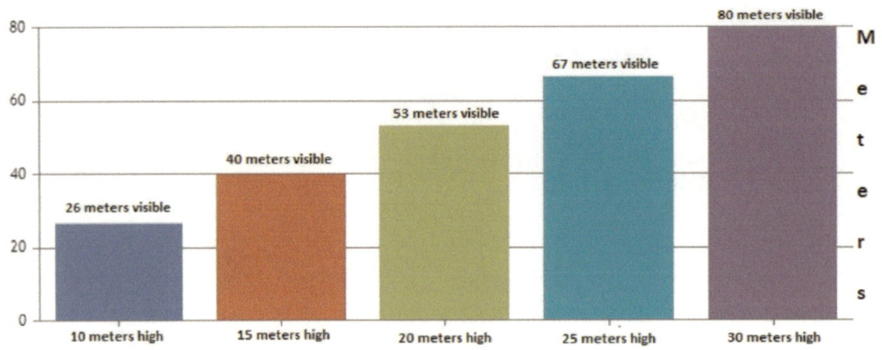

Figure 15.
Parrot Drone viewing distances in meters.

4.1 Phase 1. Distance estimation

To validate the distances shown in **Figure 15** and their visibility range, four consecutive objects are placed on the crack in the road. From **Figure 16**, only three elements can be observed which are enclosed in circles as can be seen. The dimensions

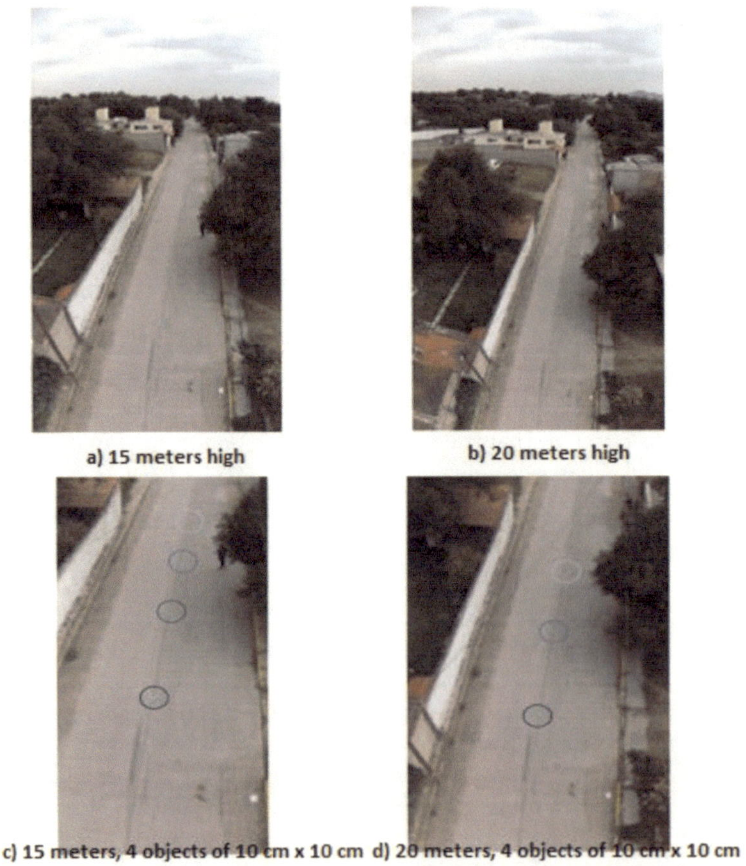

Figure 16.
Range of visibility of objects a) and b) no objects are present; c) 15 meters away, and d) 30 meters away.

of the objects placed on the crack are 10 × 10 cm, which were used to estimate the field of view of the drone camera. Based on these tests, a height of 15 meters is proposed for clear detection of the object by the drone, coupled with its stability in the air currents present in the tests.

4.2 Phase 2. Estimation of the pre-processing stage

Table 1 shows the average times calculated for the number of samples acquired, as the DWT-H decomposition increases, the average processing time increases. From **Table 1**, we observe the processing times for the feature extraction and classification stage for each decomposition scale of the DWT-H. The dimension of the initial image is 2048 × 2048, we observe that decomposition level 4 it gives a processing time of 14,445 ms for the two proposed stages.

4.3 Phase 3. Descriptors

Table 2 shows the average results obtained for the 500 images in the feature extraction stage. From **Table 1**, it is concluded that the optimal wavelet decomposition size for this estimation is at the fourth wavelet decomposition level.

4.4 Phase 4. Classification of images

The tests to validate the proposed methodology were carried out with 150 images acquired at a height of 15 meters. During the development of the test scenarios, four cases were considered: two correct classifications and two wrong classifications. The correct classifications are true positive (TP) and false positive (FP); and the misclassifications are false negative (FN) and true negative (TN). By using these metrics, we can obtain different performance measures like [13].

$$Sp = \frac{TN}{(TN + FP)} \tag{12}$$

Scale number (DWT-H)	Dimension (pixels)	Processing time average (ms)	Convolution time process average(ms)	Total processing time average (ms)
1	2048 × 2048	14.487	0.0196	14.507
2	1024 × 1024	14.506	0.0094	14.516
3	512 × 512	14.561	0.0087	14.570
4	256 × 256	14.737	0.0083	14.574

Table 1.
Result of pre-processing stage time.

Image size (pixels)	Feature extraction (statistical moments)	Feature extraction (SIFT)	Total processing time average
1024 × 1024	0.0898 (ms)	0.1826 (ms)	0.2724 (ms)

Table 2.
Descriptor stage processing time result.

$$Se = \frac{TP}{(TP + FN)} \tag{13}$$

$$Acc = \frac{(TP + TN)}{cracks\ detected\ in\ the\ image} \tag{14}$$

		Prediction	
		Positive	**Negative**
150 images	True	(TP) = 140	(FN) = 1
	False	(FP) = 5	(TN) = 4

Table 3.
Confusion matrix.

Metric	Result
Acc	0.9929
Sp	0.9655
Se	0.8000

Table 4.
The obtained results from the acquired images.

Figure 17.
Results obtained by the proposed methodology: a), c), and e) original image, b) and d) processed image (ZZC), finally f) processed image (LC).

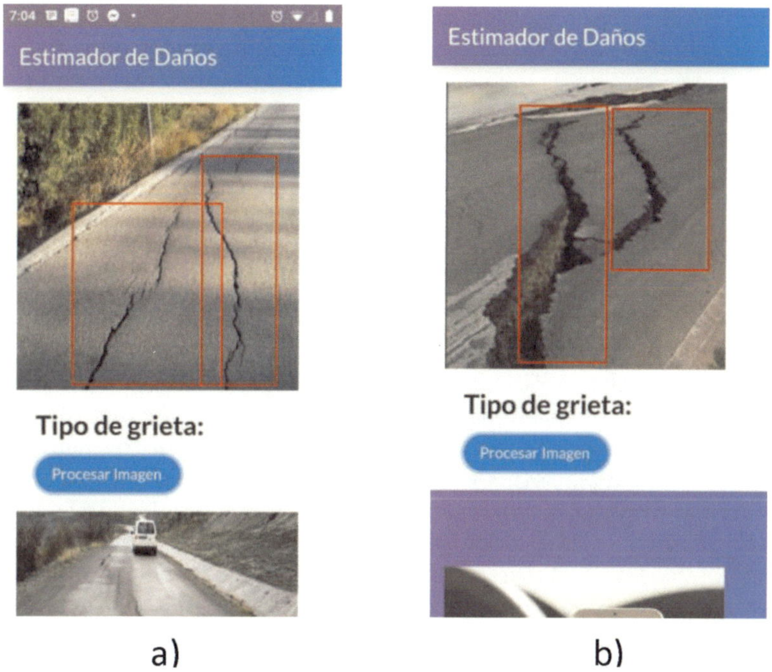

Figure 18.
Graphical user interface for interaction with the proposed methodology: a) LC detected and b) ZZC detected.

where specificity (Sp) is the ability to detect non-crack pixels, sensitivity (Se) reflects the ability of the algorithm to detect the edge of the crack, Accuracy (Acc) measures the proportion of the total number of pixels obtained correctly (sum of true positives and true negatives) by the total number of pixels that constitute the image of the cracks [13]; this is the probability that a pixel belonging to the crack image will be correctly identified. **Table 3** shows the results obtained from 150 test images that were acquired in flight, obtaining a total of 140 images (TP), 1 (FN), 5 (FP), and 4 (TN).

In **Table 4**, the results obtained from Acc, Sp, Se of the 150 acquired test images are shown. From **Table 4**, the results obtained show that 99.29% was obtained for Acc, which indicates that in this percentage the cracks were detected and classified positively. In addition, 96.55% of Sp represents that the result of no crack is true, as well as the value of Se with 80% to detect that it is not a crack.

In **Figure 17**, some images obtained through the proposed methodology and the result obtained from the classification are shown.

Finally, the mobile application that serves as the development and user interface is shown in **Figure 18**.

5. Discussion

Based on the tests carried out in the monitoring of the roads using the Parrot drone, we observed that the height between 15 and 20 meters gives satisfactory results. Within the development of the proposal, the size reduction stage made it possible to speed up the processing of the extraction of the characteristics, as well as the proposal to reduce the key points obtained by the statistical descriptors and SIFT, through Eq. (6). These development stages are fundamental because all the crack detection and identification processing are carried out internally on a mid-range mobile device. The section of the LeNet neural network was streamlined through the preprocessing stage, observing that the precision results obtained were not affected, which are 99%, even limiting the data that are entered into the neural network.

6. Conclusion

In conclusion, it can be emphasized the fact that the objectives that were sought to be achieved with the identification of cracks in roads, streets, highways or avenues, were achieved. It had the specific characteristics that allowed using the proposed processes on a mobile device, and it was possible to demonstrate that the processing of the proposed methodology was developed on an Android platform that to date is one of the most commercial platforms worldwide between mobile devices. The pre-processing results show a clear trend in terms of the time required to adapt an image and perform the crack identification process, time that does not exceed 14.79 ms, thanks to the use of DWT-H instead of other processes that require greater computational complexity for image size reduction. On the other hand, the results show that the proposed operations are 99% accurate in finding cracks. It was also found that the times of certain stages of the process can be improved by changing some processes such as the scaling of the images, which reduces the time by up to 200 milliseconds, among other possible improvements that can be implemented.

Acknowledgements

The work team is grateful for the support provided to perform this research to the Secretaria de Educación, Ciencia, Tecnología e Innovación de la Ciudad de México with the project SECITI/072/2016 and SECTEI/226/2019. We also thank the Instituto Politécnico Nacional for the research project SIP 20210178.

Conflict of interest

The authors declare no conflict of interest.

Author details

Izyalith E. Álvarez-Cisneros[1], Blanca E. Carvajal-Gámez[2*], David Araujo-Díaz[1], Miguel A. Castillo-Martínez[3] and L. Méndez-Segundo[1]

1 SEPI-ESCOM, Instituto Politécnico Nacional, Unidad Profesional Adolfo López Mateos, Ciudad de México, México

2 SEPI-UPIITA, Instituto Politécnico Nacional, La Laguna Ticoman, Gustavo A. Madero, Ciudad de México, México

3 SEPI-ESIME Culhuacan, Instituto Politécnico Nacional, San Francisco Culhuacan, Culhuacan CTM V, Coyoacán, Ciudad de México, México

*Address all correspondence to: drabecarvajalg@gmail.com

References

[1] Economic Commission for Latin America and the Caribbean. Local Economic Development and Decentralization in Latin America: Comparative Analysis. Economic Commission for Latin America and the Caribbean. 2001. Available from: https://www.cepal.org/es/publicaciones/2691-desarrollo-economico-local-descentralizacion-america-latina-analisis-comparativo [Accessed: 27 July 2021]

[2] Mexican Geological Service. Evolution of Tectonics in Mexico. Mexican Geological Service. 2017. Available from: https://www.sgm.gob.mx/Web/MuseoVirtual/Riesgos-geologicos/Evolucion-tectonica-Mexico.html [Accessed: 27 July 2021]

[3] Alden A. The World's Major Earthquake Zones. 2020. Available from: https://www.thoughtco.com/seismic-hazard-maps-of-the-world-1441205 [Accessed: 06 August 2021]

[4] Geo-México. Geo-México the Geography and Dinamics of Modern México. 2015. Available from: https://geo-mexico.com/?p=12955 [Accessed: 27 July 2021]

[5] Secretary of Communications and Transportation. Federal Roads and Bridges Package for Income and Related Services. Secretary of Communications and Transportation. 2018. Available from: https://www.sct.gob.mx/fileadmin/Transparencia/rendicion-decuentas/MD/34_MD.pdf [Accessed: 27 July 2021]

[6] Porras Díaz H, Castañeda Pinzón E, Sanabria Echeverry D, Medina Pérez G. Detección automática de grietas de pavimento asfáltico aplicando características geométricas y descriptores de forma. Dialnet. 2012;**8**:261-280

[7] Radopoulou S, Brilakis I. Patch detection for pavement assessment. Automation in Construction. 2015;**53**: 95-104. DOI: 10.1016/j.autcon.2015.03.010

[8] Tedeschi A, Benedetto F. A real-time automatic pavement crack and pothole recognition system for mobile Android-based devices. Advanced Engineering Informatics. 2017;**32**:11-25. DOI: 10.1016/j.aei.2016.12.004

[9] Maeda H, Sekimoto Y, Seto T, Kashiyama T, Omata H. Road damage detection using deep neural networks with images captured through a smartphone. Computer-Aided Civil and Infrastructure Engineering. 2018;**2018**: 1-14. DOI: 10.1111/mice.12387

[10] Zhang B, Liu X. Intelligent pavement damage monitoring research in China. IEEE Access. 2019;7: 45891-45897. DOI: 10.1109/ACCESS.2019.2905845

[11] Bhairannawar S. Efficient medical image enhancement technique using transform HSV space and adaptive histogram equalization. In: Soft Computing Based Medical Image Analysis. EEUU: Science Direct; 2018. pp. 51-60. DOI: 10.1016/B978-0-12-813087-2.00003-8

[12] Prabha D, Kumar J. Assessment of banana fruit maturity by image processing technique. Journal of Food Science and Technology. 2013;**2013**:1-13. DOI: 10.1007/s13197-013-1188-3

[13] Ramos-Arredondo RI, Carvajal-Gámez BE, Gendron D, Gallegos-Funes FJ, Mújica-Vargas D, Rosas-Fernández JB. PhotoId-Whale: Blue whale dorsal fin classification for mobile devices. PLoS One. 2020;**15**(10):

e0237570. DOI: 10.1371/journal.pone.0237570

[14] Pymasearch. LeNet – Convolutional Neural Network in Python. 2016. Available from: https://www.pyimagesearch.com/2016/08/01/lenet-convolutional-neural-network-in-python/ [Accessed: 27 July 2021]

[15] Secretary of Communications and Transport. Catalog of Deterioration in Flexible Pavements of Mexican Highways. Secretary of Communications and Transport. 1991. Available from: https://imt.mx/archivos/Publicaciones/PublicacionTecnica/pt21.pdf [Accessed: 27 July 2021]

Section 3

Sensors

Chapter 8

Adsorption-Semiconductor Sensor Based on Nanosized SnO_2 for Early Warning of Indoor Fires

Nelli Maksymovych, Ludmila Oleksenko and George Fedorenko

Abstract

The paper is devoted for a solution of indoors fires prevention at early stage by determination of H_2 (fire precursor gas) in air using a semiconductor sensor. A material based on Pt-containing nanosized tin dioxide with an average particle size of 10–11 nm obtained via a sol–gel method was created for a gas sensitive layer of the sensor. The developed sensor has high sensitivity to H_2 micro concentration, a wide range of its detectable content in air, selectivity of H_2 measuring in the presence of CO and CH_4, good dynamic properties. The combination of these properties is very important for prevention of inflammations on their early stages before the open fires appearance. Economic benefit of the proposed sensor is due to a lower cost and higher reliability of the fire situation detection.

Keywords: early fire detection, sensor, semiconductor, nanomaterial, tin dioxide, platinum

1. Introduction

Fires detection is an urgent task today, since every year, a large number of the fires occurs all over the world, leading to human casualties and economic losses. It can be seen from the **Figure 1** that fires take place in all countries of the world regardless gross domestic product values and level of technology development.

These fires appear from various sources of ignition (structure fires, vehicles, forests, grass, rubbish) as shown in **Figure 2**.

The fires that occur in premises (the structure fires) are of particular importance among all fires, since they lead to human casualties (**Figure 3**). Besides, these fires lead to significant and often irreparable economic losses.

The decrease in the number of the fires and their consequences are determined mainly by the presence of special warning means, the so-called fire-alarm-systems. When the fire-alarm-system is creating the main responsibility lies on the choice of reliable detectors appropriate for providing a timely alert of inflammation. Modern fire detectors can be conventionally divided into the following groups: **heat detectors** registering a temperature increase while the inflammation is present; **smoke detectors** operating due to ionization or photoelectric effects; **flame detectors** based

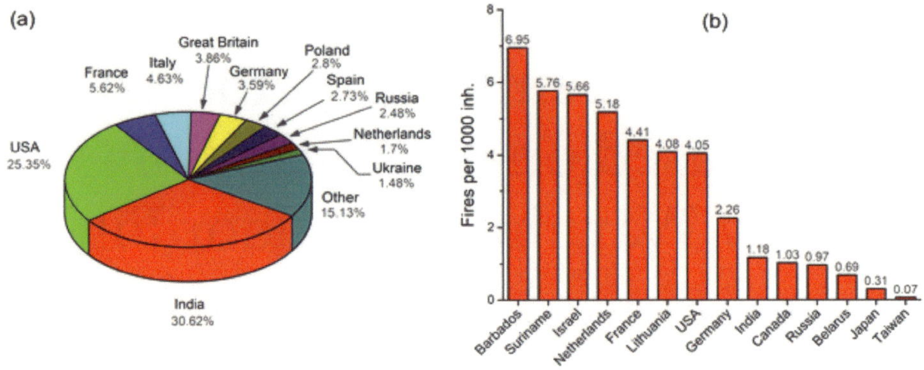

Figure 1.
Average annual fires distribution in percent within 65 countries (a) and average annual number of fires per 1000 inhabitants (b) in 2014–2018 years [1].

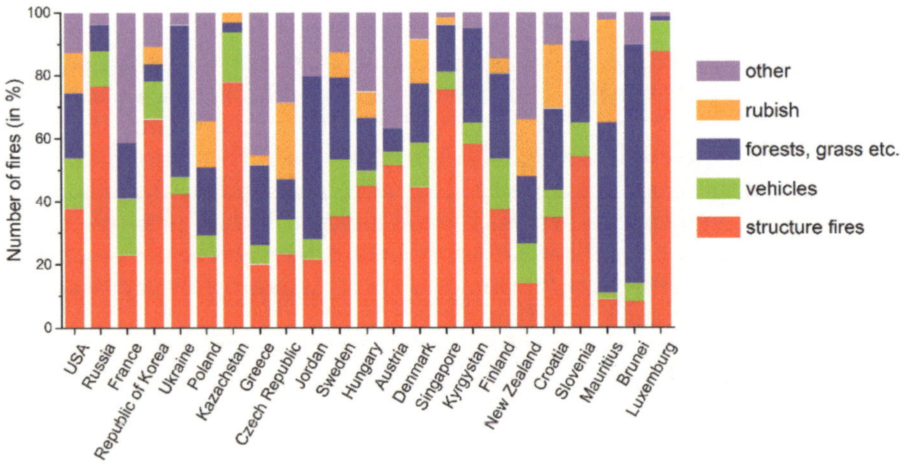

Figure 2.
Distribution of fires by types in the countries of the World in 2018 [1].

on the use of ultraviolet or infrared irradiation and **gas detectors** registering changes in gas composition of surrounded air as a result of the inflammation [2, 3].

Among them the heat detectors and flame detectors are not capable of early registration of the fire, because they response on increasing temperature caused by a hot air flow, or by electromagnetic irradiation in various ranges of a spectrum attributed to burning process. All those physical phenomena are observed when the fire is already spread around several square meters. In the contrast to the heat detectors, smoke detectors give a fire-alert before the open flame will be present. But the walls of a smoke detector chamber become dusty with time that can cause false fire-alert and heat detectors that register an increase temperature of air in rooms are rather crude and cannot detect the occurrence of the fire at the smoldering stage. **Figure 4** shows some types of the fire-alarm systems with the above mentioned detectors.

Nowadays the above mentioned kinds of the fire detectors are widely eliminated by more universal systems based on analysis of a chemical composition of air which changes dramatically when processes of thermal decomposition of the overheated inflammable materials take place [4–6]. The type of gases produced at the initial

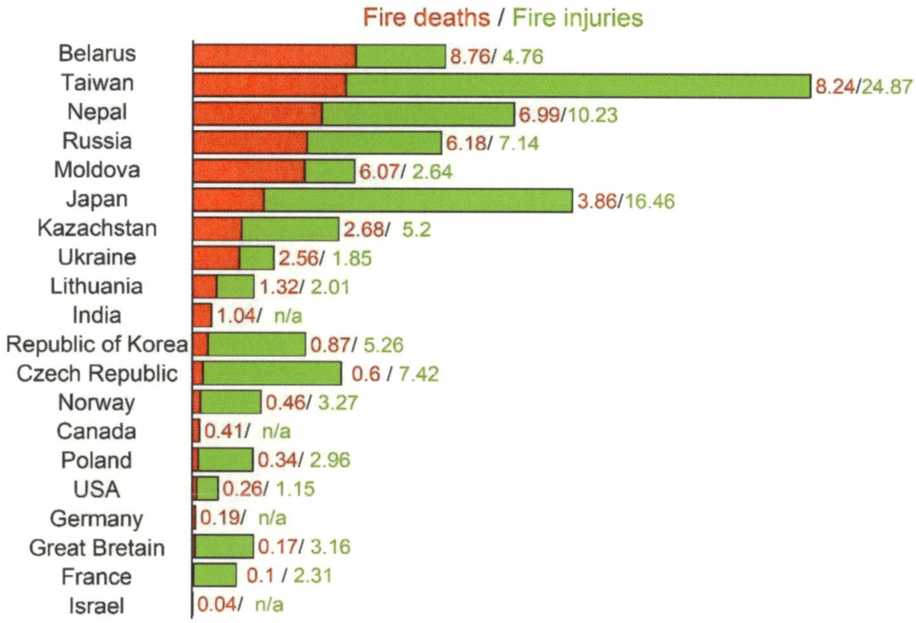

Figure 3.
Average annual number of fire death and injuries per 100 fires (in 2014–2018 years) [1].

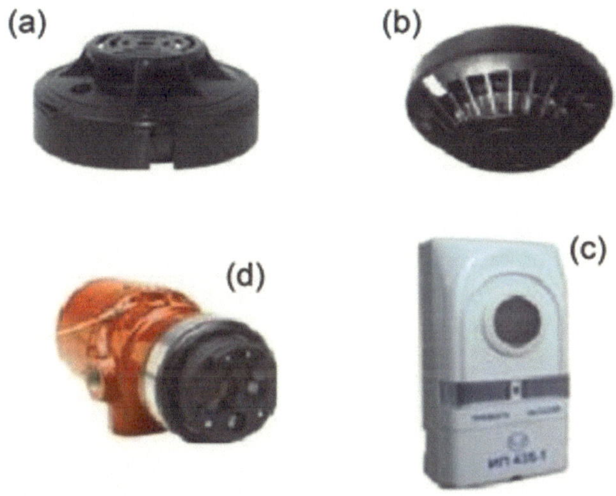

Figure 4.
Fire alarm systems based on different types of detectors: a – heat-type IP 105-1 D; b – smoke-type DUR-40 Ex; c – flame- type IP 330/1-20; d -gas type IP 435-1.

stage of combustion are determined by the composition of materials involved in this process, however, in the most cases, the main components of the appeared gases (fire precursors) can be distinguished. In particular, in the work [7] it has been experimentally established that the first gas component released during decay (pyrolysis) of wood, textiles, synthetic materials covering metal wires (telephone and electric) is hydrogen and its concentration at the early stage of ignition reaches 20–25 ppm. It is important, that appearance of CO and CH_4, during the ignition, for example, of wood, polyethylene (PE), polyurethane (PUR) and mixtures of these materials (mixed crib) is observed only at a later stage, as can be seen from the data presented in **Table 1**.

In this case, as established experimentally in study of the combustion of various objects commonly used in domestic premises (**Table 2**), the appearance of CO for some of them was not observed at all (for sponge and food), or CO appeared in air only at significant burning time (for wood). The appearance of smoke, in the amount required for responses of the smoke detectors, is observed only at a significant heating time of burning objects (except of the sponge). But the production of hydrogen was always observed for all studied materials immediately after beginning of the heat (except for the roll), and a signal values for the gas sensor reached a large value, especially for the wood and sponge (the latter used as an example of organic matter because a lot of household items in premises consist it (wallpaper, furniture, dishes, etc.)).

Thus, the detection of fire at the early stage of ignition is possible by measuring the concentration of hydrogen, which is almost always can be detected in air of the room where the fire occurs, and which appears the first of the gases - precursors of

Gas	Wood		PUR		PVC		Mixed crib	
Hydrogen	Time (min)	Signal (mV)	Time (min)	Signal (mV)	Time (min)	Signal (mV)	Time (min)	Signal (mV)
	0–10	0–15	0–5	0	0–5	<5	0–10	<5
	10–20	15–37	5–10	0–10	5–10	5–10	10–20	5–10
	20–45	37–40	10–20	10–30	10–80	10–25	20–40	10–15
	45–50	30–40	20–70	30–35			40–60	15–20
							60–80	20–30
							80–100	30
Carbon monoxide	Time (min)	Conc. (ppm)	Time (min)	Conc. (ppm)	Time (min)	Conc. (ppm)	Time (min)	Conc. (ppm)
	0–10	0–5	0–5	0	0–80	<2	0–70	<2
	10–20	5–12	5–10	0–12			70–100	2–5
	20–30	12–20	10–15	12–7				
	30–40	20–25	15–65	7				
	40–50	25–27	65–70	<7				
Smoke detectors	Time (min)	Detector alarm	Time (min)	Detector alarm	Time (min)	Detector alarm	Time (min)	Detector alarm
	5	1st detector	No alarms		No alarms		No alarms	
	10	2nd detector						

CO concentration was measured by Fourier transformed infrared spectroscopy, H_2 concentration was measured using a gas sensor based on a metal/solid electrolyte/insulator. Presence of smoke was determined by a commercial smoke detector [7].

Table 1.
Occurrence of hydrogen, carbon monoxide and smoke in air with the pyrolysis time of different materials: Wood, polyethylene (PE), polyurethane (PUR) and mixtures of these materials (mixed crib).

Gas	Armchair		Roll		Sponge		Food	
Hydrogen	Time (min)	Signal (mV)	Time (min)	Signal (mV)	Time (min)	Signal (mV)	Time (min)	Signal (mV)
	0–10	<10	0–40	<5	0–3	0–50	0–10	<5
	10–20	10–30	40–60	5–10	3–10	50–60	10–20	5–20
	20–30	30–40	60–80	10–70	10–25	60–65	20–30	20–30
	30–40	40–60	80–100	70	25–40	50–65	30–40	30
	40–45	60–70						
	45–50	50–60						
Carbon monoxide	Time (min)	Conc. (ppm)	Time (min)	Conc. (ppm)	Time (min)	Conc. (ppm)	Time (min)	Conc. (ppm)
	0–5	0	0–50	0	0–40	<5	0–40	<5
	5–10	<5	50–60	<5				
	10–20	5–10	60–70	5–40				
	20–30	10–15	70–100	40–50				
	30–40	15–25						
	40–50	25–30						
Smoke detectors	Time (min)	Detector alarm	Time (min)	Detector alarm	Time (min)	Detector alarm	Time (min)	Detector alarm
	20	1st detector	60	1st detector	2	1st detector	24	1st detector
	40	2nd detector	65	2nd detector	4	2nd detector	2nd detector – no alarm	

**CO concentration was measured by Fourier transformed infrared spectroscopy, H_2 concentration was measured using a gas sensor based on a metal/solid electrolyte/insulator. Presence of smoke was determined by a commercial smoke detector [7].*

Table 2.
*Occurrence of hydrogen, carbon monoxide and smoke in air with the time of pyrolysis of various household items: Armchair, roll, sponge, food with the time of their heating.**

the fire. At the early stage of the ignition it is still possible to assume adequate action and take control over the situation. For example, in a case of electricity wires being overheated it is not late to switch them off automatically, thus preventing the fire at the beginning.

Although the appearance of certain gases in air during the ignition of materials is separated in time, but reliable detection of the fire based on the hydrogen released at the first stage of combustion is possible only with sufficient selectivity of the sensor to hydrogen. But the sensors are known to do not possess good selectivity [8, 9].

That is why, for the early prevention of the fire, the gas sensors are included in complex systems, for example, in the MAGIC.SENS Automatic LSN Fire Detector (Bosch), that provides multipurpose detection of the fire-hazard using a joint action of optical, thermal and chemical sensors. Undoubtedly, a signal of such kind of the multiple-purpose detectors considerably increases veracity of the fire diagnostics, while the separate use of the detectors of this system do not provide effective diagnostics of the fire situation. Indeed, using only a thermal detector will not allow determining the onset of smoldering materials, in the presence of dust the false fire alerts of an optical detector are possible, and using only a gas sensor in the system will not provide diagnostics due to its low selectivity to gases emitted during the fire.

At the same time, possessing, in general, insufficient selectivity, the semiconductor gas sensors have a number of other capabilities (high sensitivities to gases, good response time, a wide range of gas detection, an ability to operate in a significant range of the ambient temperatures (-45 ÷ + 45°C), low mass and dimensions), which make them promising for creation of devices and systems capable to provide reliable diagnostics of the onset of the fires.

The design of the semiconductor sensor created at the Department of Chemistry of the Taras Shevchenko National University of Kyiv is present in **Figure 5**.

The sensor consists of a ceramic plate with platinum heater on one side and Pt contacts on the opposite side of the plate (**Figure 5a**). These contacts intended to measure the current flowing through a gas sensitive layer deposited between them (**Figure 5b**). The gas sensitive layer is made of semiconductor material, whole sensor has a very small size (2x2x0,5 mm, **Figure 5c**).

A principle of the semiconductor sensor operation is based on a change in its conductivity in the presence of the analyzed gas in air. **Figure 6** shows schematically the sensor sensitivity mechanism from the standpoint of the band theory of semiconductors [11].

In the absence of the analyzed gas in air (**Figure 6A**), chemisorption of oxygen occurs on the sensor surface with localization of electrons from the semiconductor conduction band on the oxygen atoms [12, 13]. Reduction of the electrons number in the conduction band of the semiconductor leads to decrease in its electrical conductivity. In the presence of the analyzed gas R (**Figure 6B**), its molecules interact with the active chemisorbed oxygen. The electrons, previously localized on the chemisorbed oxygen, return to the conduction band of the semiconductor and increase its conductivity. This increase will be the greater if the concentration of gas R will be higher. This is the basis for the use of semiconductor materials in the sensors for determination of the concentration of the analyzed gas R. Finally, **Figure 6C** shows a role of the catalytic additive (Me), that increases a rate of catalytic oxidation reaction by the chemisorbed oxygen through activation of the R molecules, that leads to an even larger number of electrons returned to the conduction band - the sensor

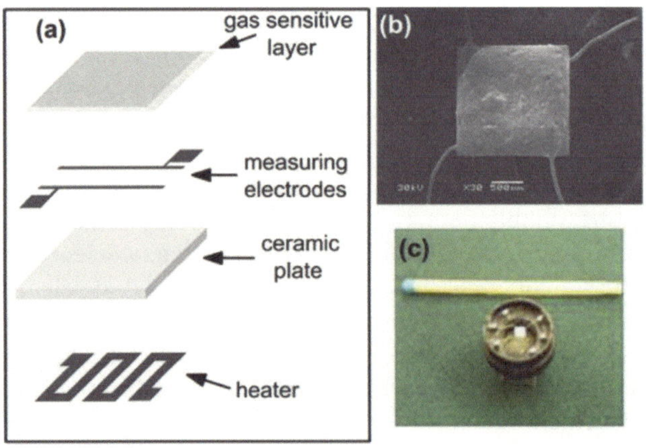

Figure 5.
Design of the semiconductor gas sensor: a – schematic illustration of the main elements of the semiconductor gas sensor; b – view of a ceramic sensor plate with a sensitive layer; c – SEM image of the sensor in its measurement chamber [10].

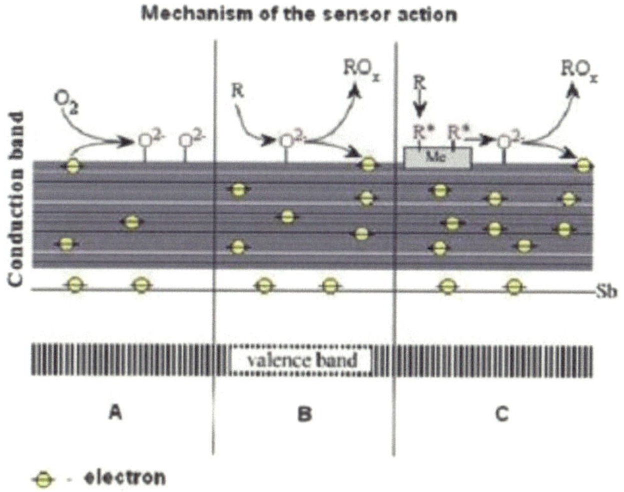

Figure 6.
Schematic illustration of the action mechanism of the semiconductor sensors: A – Sensor material in absence of analyzed gas (R); B – Sensor material exposed to the analyzed gas; C – Doped by catalytic additive me sensor material exposed to the analyzed gas.

sensitivity to the analyzed gas becomes greater [14, 15]. This simple scheme shows a reason for the low selectivity of the sensor: at a given sensor temperature, molecules of many gases can overcome the energy barrier of the oxidation reaction [16] thereby increase the conductivity of the sensor.

Several devices were created at the University using semiconductor sensors (**Figure 7**). In particular, with combination of the sensor and previous chromatographic separation of the air sample, it was possible to ensure absolute selectivity of CH_4, C_2H_6 and C_3H_8 measurements that was applied in a developed portable chromatograph intended to determine natural gas leaks from pipelines without digging the soil (**Figure 7b**).

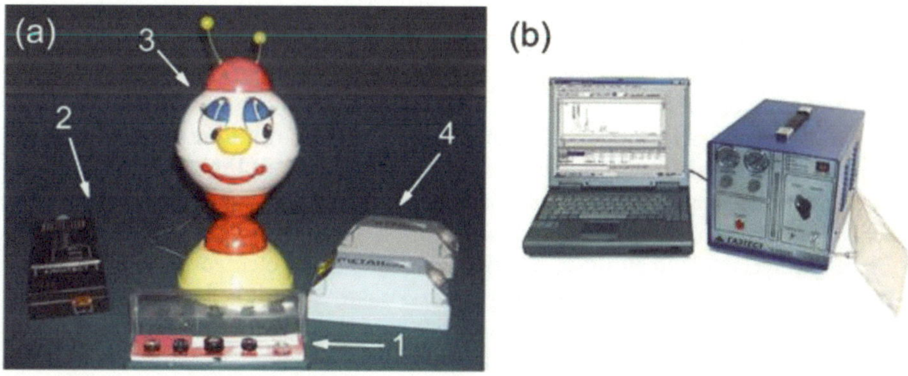

Figure 7.
Examples of developed devices based on semiconductor sensors created at the Taras Shevchenko National University of Kyiv: Semiconductor sensors (1, a), a device intended to determine alcohol in exhaled air (2,a), a domestic device in the form of a doll intended to determine presence of methane in air of kitchen (3,a), a device intended to determine concentrations of methane in air purposed for vehicles (4, a) a portable chromatograph intended to determine natural gas leaks from pipelines without digging soil (b) (Project #840 of Science and Technology Center in Ukraine).

It should be noted that although the sensors have low selectivity, changes in the chemical compositions of the gas sensitive layers, their morphology and temperature can affect the rates of gases oxidation on the semiconductor surfaces and, thus, regulate the selectivity of measurement. These approaches allow to avoid the use of the sensors in conditions of the chromatography, that neutralize attractive possibilities of using the low-power sensors in the corresponding miniature portable devices or as sensors of stationary automated systems for monitoring indoor air.

In our work, tin dioxide, an n-type semiconductor, was chosen as a basis of the sensitive layer of the sensor [17]. Tin dioxide exhibits high chemical and thermal inertness, that is necessary to ensure a long-term stable operation of the sensor in the practice [12]. In order to increase the sensor response, the material of the gas sensitive layer based on tin dioxide was obtained in a nanoscale state [13, 14, 17]. This should ensure that the processes occurring on the gas sensitive surface of the sensor have a predominant effect on its volumetric characteristics, in particular, on the values of its electrical conductivities [18]. Taking into account the mechanism of the sensor response formation [12–15, 17], a catalytically active additive was introduced into its gas sensitive layer, that allows to accelerate the process of the oxidation of hydrogen by the oxygen chemisorbed on the sensor surface. One of such catalytic additive may be platinum, which has an extremely high activity in the oxidation reaction of H_2 [19] and it is much higher than in the oxidation reaction of CO and CH_4.

The aim of our work is to develop a fast semiconductor gas sensor based on nanosized SnO_2 possesses very high sensitivity to H_2 microconcentration and low sensitivities to CO and CH_4 purposed for systems of fires detection at early stage.

2. Materials and methods of investigation

The initial substances for synthesis of the material of the sensitive layer were $SnCl_4 \times 5H_2O$, ethanediol-1,2, and H_2PtCl_6.

In order to synthesize SnO_2 1.5 g of $SnCl_4{\cdot}5H_2O$ were added to 15 ml of ethanediol-1,2 under sintering. The mixture was heated to 353 K until the salts were completely dissolved. The resulting solution was transferred to a ceramic dish and then was kept at 393 K C in a sand bath until approximately 80% of the solvent (by a volume) was evaporated. A resulting dark brown viscous gel was aged in air at a room temperature for 30 minutes, then kept in an oven at 423 K for two days. The obtained brown xerogel was subjected to heat treatment with limited air access in a temperature range from 298 K to 873 K according to a special sintering program in a programmable oven "GERO" (Germany). The temperature mode of the program is shown in **Figure 8**. A light yellow nanosized SnO_2 was obtained as a result of the treatment. Semiconductor sensors were created by applying a paste, consisted of the obtained nanosized SnO_2 and an aqueous solution of carboxymethylcellulose, between the measuring electrodes of the ceramic plates of the sensors. Then the plates with applied paste were dried at 363 K impregnated with a solution of hexachloroplatinic acid ($5{,}3 \times 10^{-2}$M) and heated up to 892 K.

Study of the obtained nanomaterials by transmission electron microscopy (TEM) was performed on an electron microscope Selmi TEM - 125 K with an accelerating voltage of 100 kV and by X-ray diffraction analysis (XRD) diffractometer on a Bruker D8 Advance diffractometer with CuKα radiation.

Study of the sensor characteristics was performed in a special electric stand (**Figure 9**).

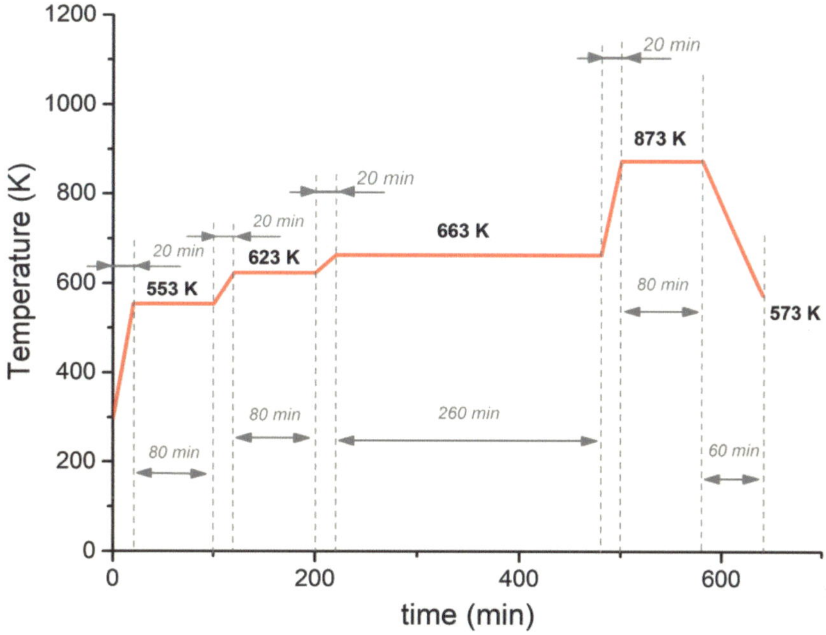

Figure 8.
A special thermal program for xerogel sintering.

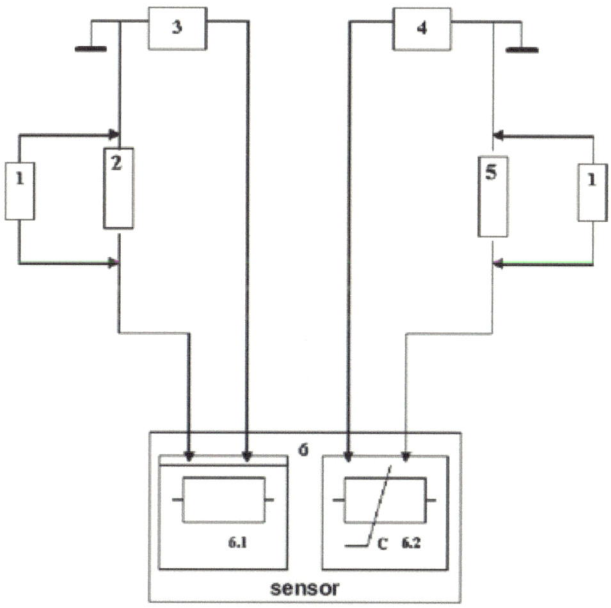

Figure 9.
Electrical diagram of the stand intended to determine the parameters of the sensors: A1 - power supply of the sensor heater; 1 – voltmeter or recording devices; 2 – a resistor for current registration through the sensor heater; 3 - power supply of the heater of the sensor; 4 - power supply of the sensitive layer of the sensor; 5 – a load resistor; 6 – a semiconductor sensor (6.1 – a heater; 6.2 – a sensitive layer).

The electrical resistance of the sensor (Rs) was calculated according to the Ohm's law for series connection of conductors:

$$Rs = R_l \left(U - U_l\right) / U_l \quad (1)$$

where U- voltage of the power supply of the sensitive layer; U_l is voltage drop on the load resistor; R_l is a value of the resistance of the load resistor.

The following notations were used to evaluate the properties of the sensors: R_o is the electrical resistance of the sensor in pure air, R_g is the electrical resistance of the sensor in the presence of an analyzed gas R. A ratio of the sensor electrical resistance in pure air to the electrical resistance in the analyzed gas – air mixture (R_o/R_g) was considered as a response of the sensor to this gas.

Before measuring the characteristics of the sensors, they were purged by a gas mixture with a hydrogen concentration of 935 ppm every 1 hour for three days at a temperature of the gas sensitive layer 675 K in order to stabilize sensors resistances.

Sensor response time ($t_{0,9}$) was defined as the time to reach 90% of the constant value of the sensor response in the presence of the analyzed mixture, the sensor relaxation time (τ_r) was defined as the time to reach the sensor 10% of the sensor response when replacing analyzed gas mixture by pure air.

The investigated gas mixtures were prepared in pressure cylinders and certified at Ukrmetrteststandard.

3. Results and discussion

Figure 10 shows a schematic illustration of nanosized tin dioxide synthesis via sol–gel method, that allows to achieve high chemical homogeneity of products, particle size control and morphology of the material at different stages of synthesis by changing temperature, reaction time, nature of solvents, chemical composition and reagent concentrations [20–22].

In contrast to the most common variant of the sol–gel method, where water is used as a solvent and the final hydroxide or oxide is obtained by alkaline or acid hydrolysis [21, 22], the non-aqueous variant of the "classical" method has a number of advantages over the method. Ethylene glycol is used as a solvent that at the same time acts as a hydrolytic agent, reacting chemically with tin (IV) chloride to form HCl

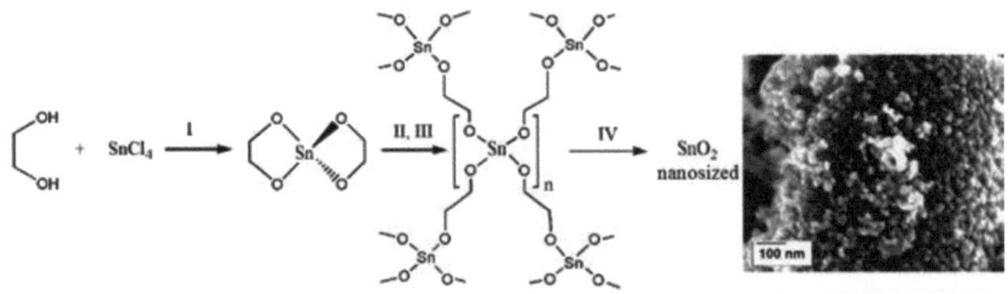

Figure 10.
Scheme of the nanosized SnO_2 synthesis by a sol–gel method: I –dissolving of $SnCl_4$ in ethylene glycol at 353 K II and III - evaporation of the obtained solution on a sand bath with consequent drying it in an oven; IV - heat treatment up to 873 K. SEM image of the synthesized SnO_2 is shown in insert.

and corresponding tin (IV) glycolate (**Figure 10**, stage I). Then, as the temperature rises, the glycolate is polymerized to form a three-dimensional grid (**Figure 10**, stages II, III), where ethylene glycol molecules act as bridges between tin atoms. Thus, ethylene glycol performs a structure-forming function. In addition, ethylene glycol also performs an important function of preventing agglomeration of particles during the heat treatment of the xerogel in an oven, as bimethylene linkers are quite stable and completely burn out of the material only at a temperature about 773 K [23]. Thus, in initial stages of the process, when the temperature rises and the oxide particles are just beginning to form, ethylene glycol molecules partially prevent the process of the nanoparticle aggregation.

According to TEM observation, the synthesized nano-SnO_2 material consists of spherical particles with an average size 10–11 nm (**Figure 11a**).

According to X-ray diffraction analysis, the obtained tin dioxide has a cassiterite structure (ICDD PDF - 2 version 2.0602 (2006), card 00. 00–041-1445), and no other phases were found in the obtained material (**Figure 12**).

Calculation of the particle size according to the Scherrer equation (XRD size) [24] has shown that the size of the coherent scattering region for the obtained material is 6.7 nm. The significant difference between the particle size according to TEM data and those calculated according to the Scherrer equation can be explained by existence of an amorphous surface layer or the presence of defects in the tin dioxide crystal lattice on the surfaces of nanoparticles [25, 26]. This may be due to the formation of the crystal structure of tin dioxide in the presence of the organic matrix, that has not yet fully oxidized in the conditions of the heat treatment [23].

It was established that the thermal formation of the semiconductor materials of the gas-sensitive layer (at temperatures up to 620°C) leads to the aggregation of SnO_2 particles. This effect can explain increase in particle size for the sensor materials in comparison with the initial SnO_2. For example, average TEM size of the semiconductor particles for the undoped sensor materials is 19–20 nm while average TEM size of the initial SnO_2 is equal to 10–11 nm [27]. No changes in the phase composition of the gas-sensitive layer were detected by the X-ray diffraction analysis, and the average particle size of tin dioxide material, calculated by the Scherrer equation, was 20.1 nm [27]. This correspondence between the values of the average particle size obtained by the different methods (TEM and X-ray diffraction) may indicate that additional high-temperature treatment in the formation of the gas sensitive materials reduces significantly a number of defects in the crystal lattice of tin dioxide.

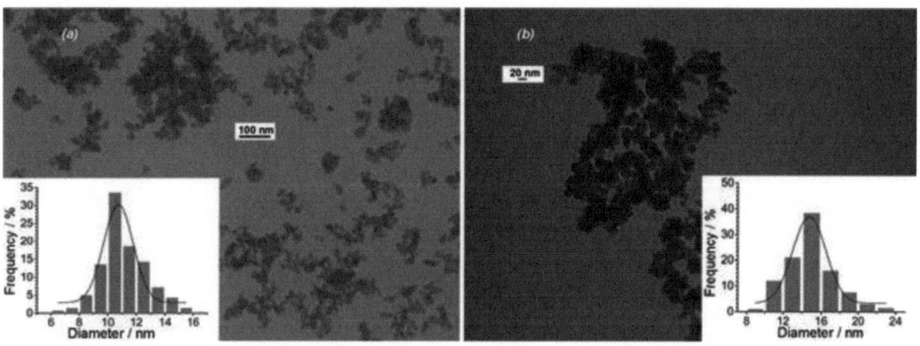

Figure 11.
TEM images of the initial tin dioxide (a) and platinum-containing sensor material obtained by impregnation with 5.3·10–2 M solution of $H_2[PtCl]_6$ (b).

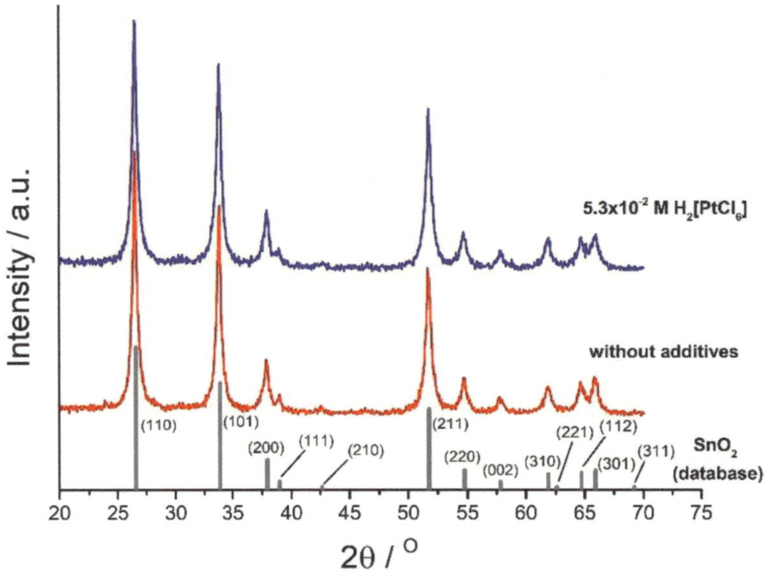

Figure 12.
X-ray diffraction of the sensor material without additives and platinum-containing sensor material (obtained by impregnation with $5.3{\cdot}10^{-2}$ M solution of $H_2[PtCl]_6$).

Analysis of TEM images of the platinum-containing gas sensitive materials (**Figure 11b**) revealed that addition of platinum reduces significantly the average particle size of the tin dioxide material (from 19 to 20 nm to 14–15 nm). The particle sizes calculated by TEM and X-ray diffraction methods are the same, that indicates to good crystallization of tin dioxide in the platinum-containing sensor material.

Figure 13a shows the dependence of the sensor response to hydrogen (22 ppm) on the temperature of the sensor. As it can be seen, this dependence has an extreme character with a maximum response of the sensor at the temperature 595 K, that may be caused by oxygen adsorption–desorption processes on the surface of the semiconductor layer.

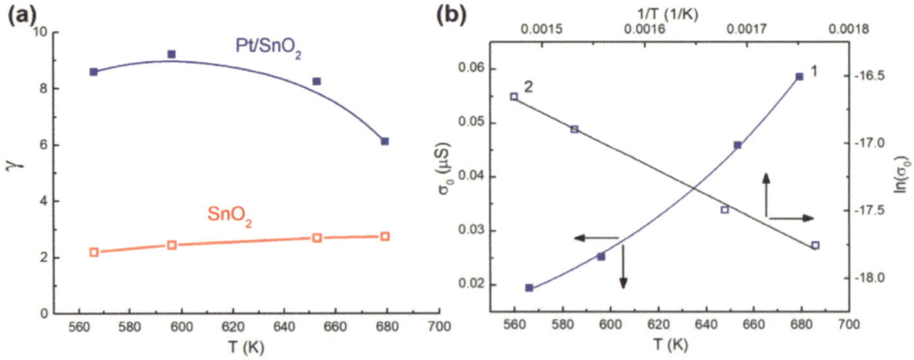

Figure 13.
The dependence of the sensor response to 22 ppm of hydrogen on the temperature of the sensor (a) and (b) - the dependence of the electrical conductivity in air (σ_0) on its temperature in the coordinates σ_0 from T (1) and ln σ_0 from 1/T (2) for the sensor based on Pt/SnO_2.

To explain the obtained data, it should be taken into account that in the presence of hydrogen the charged forms of oxygen formed by the oxygen chemisorption on the surface of the gas sensitive layer are able to oxidize H_2 molecules activated on platinum particles that leads to return of the electrons in the conduction band of the semiconductor and increases its conductivity. Naturally, if the rate of the oxidation of hydrogen is higher and, as a consequence, the greater is amount of the chemisorbed oxygen on the surface of the semiconductor is involved in oxidation of H_2 molecules, the response of the sensor will be higher. It can be assumed that increase in the operating temperature of the sensor from 565 to 595 K leads to an increase in the amount of the oxygen chemisorbed on the surface of the gas sensitive layer and consequently to an increase in the response of the sensor from 8.6 to 9.2, respectively. A further decrease in the response of the sensor with increasing the operating temperature may be due to the dominance of the desorption of the chemisorbed oxygen from the surface of the semiconductor. At the same time, this assumption is contradicted by the experimental data shown in **Figure 13b**. These data show a monotonic increase in the conductivity of the sensor under conditions of the increasing temperature, which cannot explain the initial stage of increasing the response of the sensor to hydrogen (**Figure 13a**) by increasing the amount of the chemisorbed oxygen. To understand this discrepancy, it is necessary to consider the main factors that determine it.

As can be seen from **Figure 13b**, an increase in the temperature of the sensor leads to an increase in its conductivity according to the exponential law (**Figure 13b**, curve 1), that is characteristic of semiconductors [28]. This dependence can be linearized in the coordinates ln (σ) - 1/T, that allows to calculate the activation energy of conductivity. It was found that a value of the activation energy in the operating temperature range of the sensor 565–680 K is 0.65 eV. This value is characteristic of the interparticles barrier between the grains of tin dioxide [29] indicating a slight effect of the oxygen chemisorbed at the interface between the particles of platinum and tin dioxide on the value of activation energy of the conductivity of Pt/SnO_2. That is why a change in the conductivity of the sensor from the temperature (**Figure 13b**, curve 1) does not reflect the effect of the chemisorbed oxygen on the conductivity of the sensor, which is usually observed for sensor systems [30]. Since an increase in the temperature of the sensor leads to an increase in its conductivity that is a result of an increase in the number of charge carriers (electrons) capable to overcome the interparticles barrier, an increase in the temperature leads to a decrease in the width of the space charge region (SCR). It is known that the width of the SCR is one of the factors that can directly affect the response of the sensor [12–14]. Likely, for the sensors based on nanosized Pt/SnO_2 material, an increase in the temperature leads to a significant decrease in the width of the SCR. As a result, the surface processes, such as the involving of the chemisorbed oxygen on the SnO_2 surface in the oxidation of H_2 molecules, are masked by their own high conductivity of the semiconductor and therefore the processes of the oxygen chemisorption are not reflected in the change in the conductivity of the sensor with increasing the temperature (**Figure 13b**, curve 1).

As can be seen from the data presented in **Figure 14**, the sensor created on the basis of the Pt/SnO_2 nanomaterial can measure hydrogen in air in a wide range of its concentrations (from 3 to 935 ppm).

In the logarithmic scale, the dependence of the conductivity of the sensor on the concentration of hydrogen in air becomes linear (**Figure 14**, curve 2) [31, 32]. That is very convenient to use the sensor in practice, because its periodic calibration, that usually accompanies the long-term operation of the sensor, can be carried out only by two concentrations of hydrogen-air mixtures.

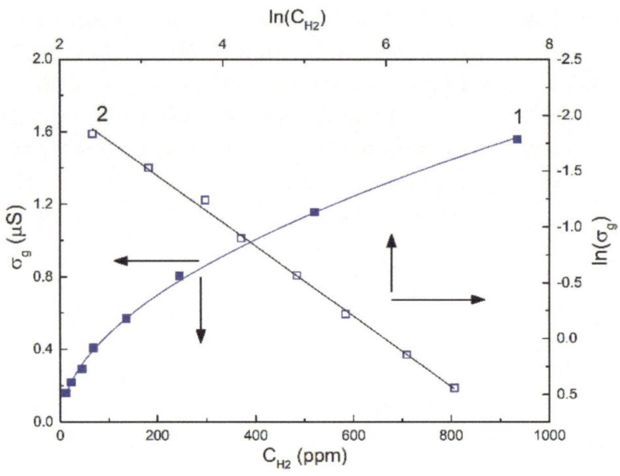

Figure 14.
Dependence of electrical conductivity of the platinum-containing sensor on hydrogen concentration in air (1) and its linearized form in the logarithmic scale (2).

One of the most important properties of sensors is the reproducibility of their performance over a period of their operation. The data presented on **Figure 15** shows a change in the conductivity of the sensor with periodic supply of 44 ppm H_2 for 100 minutes of its continuous operation. As it can be seen, the sensor signal is well reproduced -a deviation of its signal value does not exceed (± 4%) from the initial value.

Table 3 shows the data on the stability of the sensor response in the presence of 44 ppm H_2 during continuous operation for a working day (8 hours). As it can be seen, a deviation of the response value of the sensor (Δ) from its initial value (in the first hour of the sensor operation) is insignificant, that indicates to sufficient stability of the sensor.

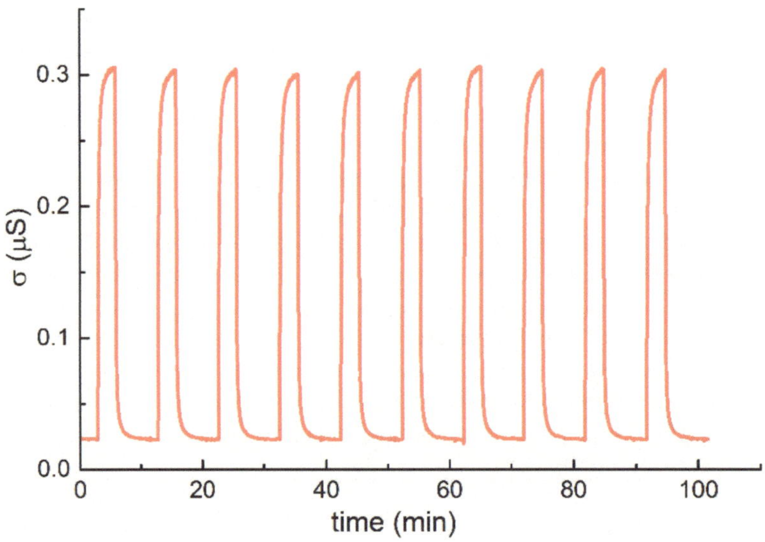

Figure 15.
The reproducibility of the platinum-containing sensor signal to 44 ppm of H_2 for 100 minutes of the sensor operation at its temperature of 595 K.

Sensor response and its change	Time (hours)							
	1	2	3	4	5	6	7	8
γ	13.22	12.95	13.412	13.21	13.66	13.19	13.24	13.59
Δ (%)		-1.98	1.45	−0.05	3.38	−0.20	0.21	2.82

Table 3.
Stability of the sensor response to 44 ppm H_2 at a sensor temperature of 595 K for 8 hours of the sensor continuous operation.

The results of measuring the sensor response to 22 ppm of hydrogen before and after overloading the sensor with hydrogen (935 ppm), are presented in **Figure 16**.

As it can be seen from the above data, the hydrogen overload has almost no effect on the value of the conductivity of the sensor, that corresponds to the presence in air of a H_2 microconcentration (22 ppm). This indicates to the stability of the material of the gas-sensitive layer, that remains unchanged when exposed to a high concentration of hydrogen at relatively high operating temperatures of the sensor. Thus, the sensor is able to correctly measure the hydrogen content even after exposure to almost 50 times the concentration of H_2 This, of course, indicates to possibility of the stable sensor operation in real conditions. In addition, these data confirm good dynamics of the sensor, because it quickly returns to its initial state in air after exposure of high hydrogen concentrations to it. In particular, it was found that at the operating temperature of 595 K, the response time of the sensor ($t_{0.9}$) in relation to 22 ppm H_2 is 10 s, and the relaxation time is 30 s.

To assess a possible effect on the conductivity of the sensor of carbon monoxide and methane, which are also classified as fire precursor gases and which appear in air of the room behind hydrogen, the same CO and H_2 concentration in air were applied to the sensor. It was found (**Figure 17**), that the conductivity of the sensor in the presence of H_2 is much higher than for CO (the value of the conductivity of the sensor for CO is about 10% of the conductivity of the sensor to H_2), and the sensor is practically not felt CH_4

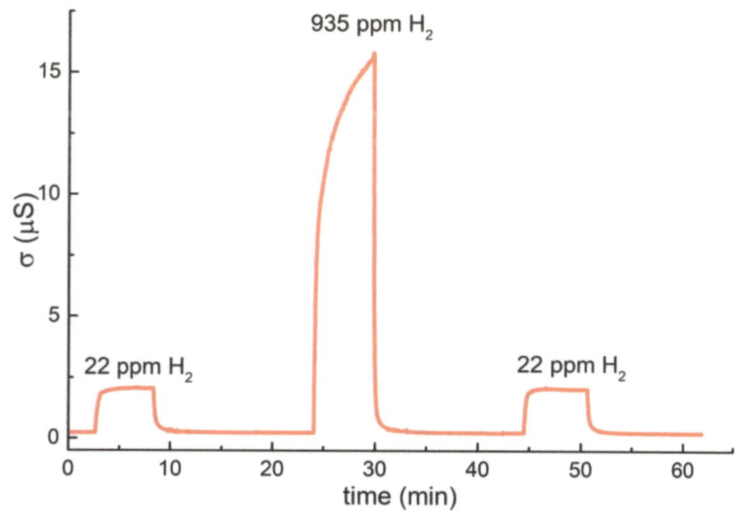

Figure 16.
Change in the conductivity of the platinum-containing sensor to 22 ppm of hydrogen after it exposure to 935 ppm of hydrogen at a sensor temperature 595 K.

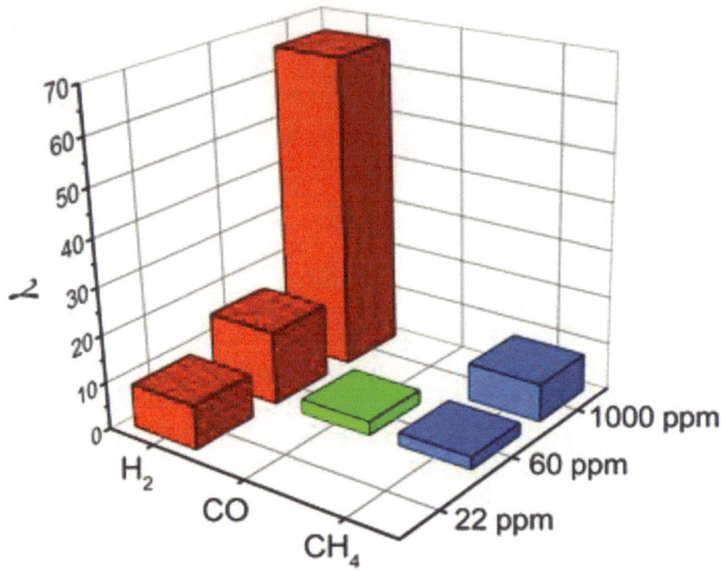

Figure 17.
The response of the Pt/SnO$_2$-based sensor to different concentrations of hydrogen (22 ppm), carbon monoxide (60 ppm) and methane (935 ppm).

at even much higher concentration (935 ppm) than for H_2. Since CO and CH_4 appear in air after some time after the appearance of hydrogen, and the sensor response time for H_2 is 10 seconds, it becomes clear that such sensor can detect the presence of H_2 before presence of CO and CH_4 in air and thus detect the earliest stage of appearance of the fire.

The proposed sensor for the fire diagnostics will have the following advantages in comparison with already known analogues: presence of the open flame will not be necessary for generating the alert (as in the case of the heat or smoke detectors), and the detector will not be impeded by the dust (as in the case of the optical sensors). The assumed high sensitivity of the sensor to H_2 (the first precursor of the fire) and the good response time will allow to determine the onset of smoldering at the earliest stage before the start of the open fire). In addition, taking into account the low power consumption of the sensors (0.4 watts), they can be used in controlled rooms with organizing their power supply from the electrical networks, that is usually presented in the rooms, which is convenient when using sensors in practice.

It should be noted that production of hydrogen is usually highest at first stage of the ignition while with appearance of the open fire production of hydrogen can drastically decrease [7, 33] while production of smoke and other organic and inorganic compounds (such as benzene, biphenyl, ethane, hydrogen chloride, hydrogen cyanide etc.) has begun [33]. This fact allows to suggest the use of the created hydrogen sensor with other type fire sensors in a complex device. This will allow to determine the initial stage of the fire situation by monitoring of the hydrogen sensor signal and to track fire development.

4. Conclusions

In this work by using a sol–gel method sensor material based on nanosized tin dioxide with an average particle size of 10–11 nm was created and possibility of its use as a gas-sensitive layer of the semiconductor hydrogen sensor was investigated. It was found that the addition of platinum to the synthesized nanomaterial increases the sensitivity of sensors to hydrogen compared to tin dioxide. It was shown that due to high sensitivity to hydrogen, fast response time, a wide range of H_2 detectable concentrations in air, resistance to hydrogen overloads, good reproducibility of the sensor signal and selectivity to hydrogen measurement, the created sensor is promising for use it in the detector of early warning of the fire.

Author details

Nelli Maksymovych*, Ludmila Oleksenko and George Fedorenko
Taras Shevchenko National University of Kyiv, Ukraine

*Address all correspondence to: nellymax@univ.kiev.ua

References

[1] Center of Fire Statistic. World Fire Statistics [Internet]. 2020. Available from: https://www.ctif.org/sites/default/files/2020-06/CTIF_Report25.pdf [Accessed: 2021-06-02]

[2] Sam Kubba. Impact of Energy and Atmosphere. In: Sam Kubba, editor. LEED v4 Practices, Certification, and Accreditation Handbook. 2nd ed. Oxford: Butterworth-Heinemann, Elsevier; 2016. p. 409-518. DOI: 10.1016/B978-0-12-803830-7.00009-8

[3] A. Enis Çetin, Bart Merci, Osman Günay, Behçet Uğur TöreyinSteven Verstockt. Methods and Techniques for Fire Detection. London: Academic Press, Elsevier; 2016. 91 p. DOI: 10.1016/C2014-0-01269-5

[4] Simone Krüger, Anka Berger, Ulrich Krause. Chemical–analytical investigation of fire products in intermediate storages of recycling materials. Fire and Materials. 2012;36:165-175. DOI: 10.1002/fam.1098

[5] Simone Krüger, Anja Hofmann, Anka Berger, Nicolas Gude. Investigation of smoke gases and temperatures during car fire – large-scale and small-scale tests and numerical investigations. Fire and Materials Fire Mater. 2016;40:785-799. DOI: 10.1002/fam.2342

[6] Ayako Sawada, Tsubasa Higashino, Takashi Oyabu, Yoshinori Takei, Hidehito Nanto, Kiyoshi Toko. Gas sensor characteristics for smoldering fire caused by a cigarette smoke. Sensors and Actuators B: Chemical. 2008;130:88-93. DOI: 10.1016/j.snb.2007.07.083

[7] Simone Krüger, Marie-Claire Despinasse, Tina Raspe, Kai Nörthemann, Werner Moritz. Early fire detection: Are hydrogen sensors able to detect pyrolysis of house hold materials? Fire Safety Journal. 2017;91:1059-1067. DOI: 10.1016/j.firesaf.2017.04.035

[8] C. Wang, L. Yin, L. Zhang, D. Xiang, R. Gao. Metal Oxide Gas Sensors: Sensitivity and Influencing Factors. Sensors. 2010;10:2088-2106. DOI: 10.3390/s100302088

[9] D. Kohl. Surface processes in the detection of reducing gases with SnO_2-based devices. Sensors and Actuators. 1989;18:71-113. DOI: 10.1016/0250-6874(89)87026-X

[10] George Fedorenko, Ludmila Oleksenko, Nelly Maksymovych, Inna Vasylenko. Cerium-doped SnO_2 nanomaterials with enhanced gas-sensitive properties for adsorption semiconductor sensors intended to detect low H_2 concentrations. Journal of Materials Science. 2020;55:16612-16624. DOI: 10.1007/s10853-020-05199-w

[11] George F. Fine, Leon M. Cavanagh, Ayo Afonja, Russell Binions. Metal Oxide Semi-Conductor Gas Sensors in Environmental Monitoring. Sensors. 2010;10:5469-5502. DOI: 10.3390/s100605469

[12] A.V. Marikutsa, M.N. Rumyantseva, A.M. Gaskov, A.M. Samoylov. Nanocrystalline Tin Dioxide: Basics in Relation with Gas Sensing Phenomena. Part I. Physical and Chemical Properties and Sensor Signal Formation. Inorganic Materials. 2015;51:1329-1347. DOI: 10.1134/S002016851513004X

[13] Noboru Yamazoe, Kengo Shimanoe. New perspectives of gas sensor technology. Sensors and Actuators B: Chemical. 2009;138:100-107. DOI: 10.1016/j.snb.2009.01.023

[14] Nicolae Barsan, Udo Weimar. Conduction model of metal oxide gas sensors. Journal of Electroceramics. 2001;7:143-167. DOI: 10.1023/A:1014405811371

[15] Ludmila Oleksenko, George Fedorenko, Nelly Maksymovych. Effect of heterogeneous catalytic methane oxidation on kinetics of conductivity response of adsorption semiconductor sensors based on Pd/SnO_2 nanomaterial. Research on Chemical Intermediates. 2019;45:4101-4111. DOI: 10.1007/s11164-019-03893-2

[16] Ašperger S. Chemical Kinetics and Reaction Mechanisms. In: Chemical Kinetics and Inorganic Reaction Mechanisms. Boston: Springer; 2003 p.3-103. DOI: 10.1007/978-1-4419-9276-5_2

[17] G. Korotcenkov, S.-D. Han, B.K. Cho, V. Brinzari. Grain Size Effects in Sensor Response of Nanostructured SnO_2- and In_2O_3-Based Conductometric Thin Film Gas Sensor. Critical Reviews in Solid State and Materials Sciences. 2009;34:1-17. DOI: 10.1080/10408430902815725

[18] Emil Roduner. Size matters: why nanomaterials are different. Chemical Society Reviews. 2006;35:583-592. DOI: 10.1039/B502142C

[19] Boris V. L'vov, Andrew K. Galwey. Catalytic oxidation of hydrogen on platinum: Thermochemical approach. Journal of Thermal Analysis and Calorimetry. 2013;112:815-822. DOI: 10.1007/s10973-012-2567-0

[20] Kalim Deshmukh, Tomáš Kovářík, Tomáš Křenek, Denitsa Docheva, Theresia Stich, Josef Pola. Recent advances and future perspectives of sol–gel derived porous bioactive glasses: a review. RSC Advances. 2020;10:33782-33835. DOI: 10.1039/D0RA04287K

[21] Mritunjaya Parashar, Vivek Kumar Shukla, Ranbir Singh. Metal oxides nanoparticles via sol–gel method: a review on synthesis, characterization and applications. Journal of Materials Science: Materials in Electronics. 2020;31:3729-3749. DOI: 10.1007/s10854-020-02994-8

[22] C. Jeffrey Brinker, George W. Scherer. Sol-Gel Science. The Physics and Chemistry of Sol–Gel Processing. San Diego: Academic Press; 1990. 908 p. DOI: 10.1016/C2009-0-22386-5

[23] E. V. Sokovykh, L. P. Oleksenko, N. P. Maksymovych, I. P. Matushko. Influence of temperature conditions of forming nanosized SnO_2-based materials on hydrogen sensor properties. Journal of Thermal Analysis and Calorimetry. 2015;121:1159-1165. DOI: 10.1007/s10973-015-4560-x

[24] C. Hammond. The basics of crystallography and diffraction. 4th ed. Oxford: Oxford university press; 2015. 507 p. DOI: 10.1093/acprof:oso/9780198738671.001.0001

[25] Holger Borchert, Elena V. Shevchenko, Aymeric Robert, Ivo Mekis, Andreas Kornowski, Gerhard Grübel, Horst Weller. Determination of Nanocrystal Sizes: A Comparison of TEM, SAXS, and XRD Studies of Highly Monodisperse $CoPt_3$ Particles. Langmuir. 2005;21:1931-1936. DOI: 10.1021/la0477183

[26] I.V. Vasylenko, S.V. Kolotilov, I.E. Kotenko, K.S. Gavrilenko, F. Tuna, G.A. Timco, R.E.P. Winpenny, V.V. Pavlishchuk. Magnetic properties of nanosized γ-Fe_2O_3 and α-$(Fe_{2/3}Cr_{1/3})_2O_3$, prepared by thermal decomposition of heterometallic single-molecular precursor. Journal of Magnetism and Magnetic Materials. 2012;324:595-601. DOI: 10.1016/j.jmmm.2011.08.049

[27] George Fedorenko, Ludmila Oleksenko, Nelly Maksymovych, Galina Skolyar, Oleksandr Ripko. Semiconductor Gas Sensors Based on Pd/SnO_2 Nanomaterials for Methane Detection in Air. Nanoscale Research Letters. 2017;12:329. DOI: 10.1186/s11671-017-2102-0

[28] Maria Cristina Carotta, Michele Benetti, Elena Ferrari, Alessio Giberti, Cesare Malagù, Marco Nagliati, Beatrice Vendemiati, Giuliano Martinelli. Basic interpretation of thick film gas sensors for atmospheric application. Sensors and Actuators B: Chemical. 2007;126:672-677. DOI: 10.1016/j.snb.2007.04.016.

[29] C. Malagù, V. Guidi, M. Stefancich, M. C. Carotta, G. Martinelli. Model for Schottky barrier and surface states in nanostructured n-type semiconductors. Journal of Applied Physics. 2002;91: 808-814. DOI: 10.1063/1.1425434

[30] A.I. Buvaylo, L.P. Oleksenko, N.P. Maksimovich, I.P. Matushko, A.P. Ripko, V.P. Ruchko. Effect of SnO_2 particle size on the hydrogen sensitivity of adsorption-semiconductor sensors with Co_xO_y/SnO_2 active coating. Theoretical and Experimental Chemistry. 2010;46:153-157. DOI: 10.1007/s11237-010-9132-3

[31] S. Roy Morrison. Mechanism of semiconductor gas sensor operation. Sensors and Actuators. 1987;11:283-287. DOI: 10.1016/0250-6874(87)80007-0

[32] Noboru Yamazoe, Kengo Shimanoe. Theory of power low for semiconductor gas sensors. Sensors and Actuators B: Chemical. 2008;128:566-573. DOI: 10.1016/j.snb.2007.07.036

[33] Ministry of Housing, Communities & Local Government of United Kingdom. Grenfell environmental checks: review of combustion related fire products. [Internet]. 2020. Available from: https://assets.publishing.service.gov.uk/government/uploads/system/uploads/attachment_data/file/928024/Combustion_related_fire_products_review_ISSUE.pdf [Accessed: 2021-06-19]